Computer-Assisted Analysis of Mixtures and Applications

Meta-analysis, disease mapping and others

MONOGRAPHS ON STATISTICS AND APPLIED PROBABILITY

General Editors

D.R. Cox, V. Isham, N. Keiding, T. Louis, N. Reid, R. Tibshirani, and H. Tong

Computer-Assisted Analysis of Mixtures and Applications

Meta-analysis, disease mapping and others

DANKMAR BÖHNING

Associate Professor
Department of Epidemiology
Institute for Social Medicine and Medical Psychology
Free University Berlin
Berlin, Germany

CHAPMAN & HALL/CRC

Boca Raton   London   New York   Washington, D.C.

Böhning, Dankmar.
    Computer-assisted analysis of mixtures and applications : meta
-analysis, disease mapping, and others / Dankmar Böhning
        p.   cm. -- (Monographs on statistics and applied probability)
    Includes bibliographical references and index.
    ISBN 0-8493-0385-0 (alk. paper)
    1. Epidemiology--Statistical methods. 2. Epidemiology--Data
processing. 3. Meta-analysis. I. Title. II. Series.
RA625.2.M3B64 1999
616.4′07′27—dc21   98-10833
                                                                          CIP

To Nat

Contents

Preface

The occurrence of mixture distributions is manifold. C.R. Rao mentions the following problem in his book (Rao 1989: p.11, *eliciting responses to sensitive questions*): a question is asked which is related to a sensitive issue such as: *Do you drink alcohol regularly? (A)*. To guarantee an anonymous procedure a second question, less sensitive, is formulated such as: *Does your residence address number end with an even figure? (B)*. Now, the interviewer is telling the person to be interviewed that only one question of the two is asked with equal chance. Which one, only the interviewed person knows. The interview only records a *yes* or *no* answer to one question *(Y)* which might be *(A)* or *(B)*. Obviously,

$$\Pr(Y = yes) = \Pr(A = yes) \; {}^1\!/_2 + \Pr(B = yes) \; {}^1\!/_2,$$

from which $\Pr(A = yes) = 2 \; \Pr(Y = yes) - \Pr(B = yes)$ can be easily calculated (Rao, 1989, p.35). Usually, the relative frequency f_B of house numbers ending with an even figure is known and the relative frequency f_Y of people answering *yes* to the question Y asked is found in the interview sample, so that the estimated relative frequency of persons drinking alcohol regularly is

$$f_A = 2f_Y - f_B.$$

The problem above involves a mixture distribution: the population consists of two components (questions) with equal proportions (equal probabilities to be asked). What is observed is the marginal distribution of Y (the question finally asked). The example has some simple features: the mixing proportions are known $({}^1\!/_2)$ as well as one component distribution $(\Pr(B = yes))$. This allows the simple solution.

Unfortunately, mixture models are more complex in full generality. In recent years some structures of mixture models could be identified which allow a beautifully simple treatment of some problems involved in mixture models. These structures are mainly connected with the

convex geometry of mixture likelihoods which lead to simple characterizations of maximum likelihood estimators and globally convergent
algorithms for their construction. This book reviews some of these
recent developments in the area of nonparametric mixture models
which have become quite popular over the last 15–20 years. One reason
is that they provide a natural framework for practitioners (the biome-
trician, the epidemiologist, the ecologist, the econometrician) to deal
with *unobserved heterogeneity*. This situation arises when under standard conditions a certain (usually simple) model is valid. However,
because of variation of the parameters describing the model in the
population, these assumptions are no longer met, though they are still
true in the subpopulations described by the variations of the parameters. Since one has not observed to which subpopulation each observed
datum belongs, one can treat the variable which is describing the
subpopulation membership only as a *latent* variable. The corresponding marginal model is a specific form of nonparametric mixture model.
The book sheds light on this approach for various important application fields such as disease mapping or meta-analysis.

A second reason for the popularity of mixture models can be seen
in the fact that mixture models provide the natural framework for
Empirical Bayes models. The cornerstone of EB-models is the estimation of the prior distribution, which can also be accomplished with
nonparametric mixture models.

In addition to developments in theory and algorithms the work
focuses on developments in applications, such as meta-analysis, disease mapping, fertility studies, estimation of prevalence under clustering, and estimation of the distribution function of survival time
under interval-censoring.

The approach is nonparametric for the mixing distribution,
including leaving the number of components (subpopulations) of the
mixing distribution unknown, one of the most attractive features of
the approach.

The computer program C.A.MAN (developed by the author) is intro-
duced and explained in the book, which will allow readers to carry out
mixture analysis of their own data in a simple manner.

Early interest in mixture models on the author's part came up in
1985–1986, when the author spent the academic year as *Visiting Associate Professor* at the Department of Statistics of the Pennsylvania
State University. The visit was arranged and initiated by Prof. Bruce
Lindsay. The author's own ideas and knowledge on mixtures have been
influenced since this period by cooperation with Bruce. A series of joint
papers indicate the fruitfulness of this cooperation.

The book was initiated by a course which the author held in December 1995 in Leuven (Belgium) at the Department of Biostatistics of the Catholic University of Leuven. The course was organized by Prof. Emmanuel Lesaffre who also encouraged the author to finally arrange the course material to become a book of its own. The course was again held in June 1998 at the Department of Statistics of the University of Florence at the invitation of Prof. Annibale Biggeri. Between these two landmarks various chapters have been changed, newly developed or reorganized at various international places and on various occasions. Chapters 2 and 3 took their final form while the author was on a leave of absence from the Free University of Berlin in the winter term 1997/98 to represent a chair in statistics at the Department of Statistics at the Ludwig-Maximilians-University in Munich. This temporary professorship was arranged by Prof. Iris Pigeot, Prof. Ludwig Fahrmeir, and Prof. Hans Schneeweiß. Chapter 5 on meta-analysis was influenced through cooperation with Prof. Heinz Holling from the Faculty of Psychology in Münster and underwent various changes during a visiting term in 1996 in Vienna at the Faculty of Psychology of the University of Vienna, which was arranged by Prof. Anton Formann. Chapter 6 on estimates of the variance of the mixing distribution was developed together with Prof. Jesus Sarol (Manila) on the occasion of a visit in March 1998 to the Phillipines. Most chapters were worked out while the author was spending summer visits from 1993–1998 at the Department of Biostatistics at Mahidol University (Bangkok, Thailand), sponsored by the German Academic Exchange Service. My academic work in Thailand would have not been possible without the help of Ajarn* Rampai, Ajarn Thavatchai, Ajarn Chukiat, Ajarn Piangchan, Ajarn Dechavudh, and so many others. Ajarn Nutkamol took the burden of introducing me into the first steps of Thai language. Joint work with Ajarn Rampai on spatial variation on the birth-ratio in Thailand went into the book as Section 1.6.

Many ideas in the book were developed jointly with my co-workers Dr. Ekkehart Dietz, Dr. Peter Schlattmann, and Dr. Uwe Malzahn in the working group of quantitative epidemiology at the Institute of Social Medicine of the Free University Berlin. In particular, Chapter 7 on disease mapping is due to joint work with Dr. Peter Schlattmann. Joint collaboration with Dr. Mathias Greiner from the Faculty of Veterinary Medicine went into the book to form Section 8.4. The head of our Institute at the Free University Berlin, Prof. Frank-Peter Schelp, has given me continuous support over the many years I have been at the Department.

---

* *Ajarn* is the expression in Thai language for a lecturer at a university.

I am sure that I have forgotten some names in the above list of contributors to the book. At least in this respect, I am sure that the list is incomplete. Nevertheless, I would like to express my deepest thanks to all my friends and colleagues – if they have been mentioned above or not – for their energy and help, encouragement and spirit, patience and attention, willingness and cooperation, humor and company. Without them this book would not exist.

Finally, I would like to thank the publisher, particularly Mark Pollard, Stephanie Harding, and Mimi Williams for their excellent performance as counterparts on the side of the publisher.

This book was written while I was receiving research support from the *German Research Foundation* (DFG), the *Ministry of Education, Science, Research and Technology* (BMBF), the *German Academic Exchange Service* (DAAD), and the European community program *BIOMED2*. I am most grateful for their support.

Dankmar Böhning
Bangkok and Berlin, July 1998

CHAPTER 1

Introduction

1.1 Population heterogeneity: the natural genesis of mixture models

The importance of *mixture distributions,* their enormous developments and their frequent applications over recent years is due to the fact that mixture models offer natural models for *unobserved population heterogeneity.* What does this mean? Suppose we are dealing with the case that a *one-parametric density* $f(x;\lambda)$ can be assumed for the phenomenon of interest. Here λ denotes the parameter of the population, whereas x is in the sample space X, a subset of the real line. We call this the *homogeneous* case and it is visualized in Figure 1.1.

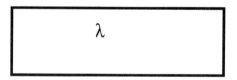

Figure 1.1. Homogeneous population

However, this model often is too strict to capture the variation of the parameter over a diversity of subpopulations. In this case, we have that the population consists of various subpopulations, denoted by λ_1, λ_2, ..., λ_k where k denotes the number (possibly unknown) of subpopulations. We call this situation the *heterogeneous case* (see Figure 1.2).

In contrast to the homogeneous case, we have the same type of density in each subpopulation j, but a potentially different parameter: $f(x;\lambda_j)$ is the density in subpopulation j.

In the sample $x_1, x_2, ..., x_n$ it is not observed, however, from which subpopulation the observations are coming. Therefore, we speak of *unobserved heterogeneity* or *extra-population heterogeneity.* Let a latent

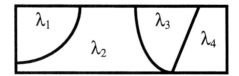

Figure 1.2. Heterogeneous case

variable Z describe the population membership. Then the joint density $f(x, z)$ can be written as $f(x, z) = f(x|z)p(z) = f(x;\lambda_z)p_z$, where $f(x|z) = f(x;\lambda_z)$ is the density conditional on membership of subpopulation z. Therefore, the unconditional density $f(x)$ is the *marginal density*

$$f(x, P) = \sum_{z=1}^{k} f(x|z)p(z) = \sum_{j=1}^{k} f(x;\lambda_j)p_j \qquad (1.1)$$

where the margin is taken over the latent variable Z. Note that p_j is the probability of belonging to the jth subpopulation having parameter λ_j. Therefore, the p_j have to meet the constraints $p_j \geq 0$, $p_1 + \dots + p_k = 1$. Note that (1.1) is a mixture distribution with *kernel* $f(x, \lambda)$ and *mixing distribution* $P = \begin{pmatrix} \lambda_1 \dots \lambda_k \\ p_1 \dots p_k \end{pmatrix}$ in which weights $p_1, \dots, p_k$ are given to parameters $\lambda_1, \dots \lambda_k$. Estimation is done conventionally by maximum likelihood, that is we have to find $\hat{P}$ which maximizes the log-likelihood $l(P) = \sum_{i=1}^{n} \log f(x_i, P)$. $\hat{P}$ is called the *nonparametric maximum likelihood estimator* (Laird 1978). The software package C.A.MAN (Böhning, Schlattmann and Lindsay 1992) provides the NPMLE for P:

$$\hat{P} = \begin{pmatrix} \hat{\lambda}_1 \dots \hat{\lambda}_{\hat{k}} \\ \hat{p}_1 \dots \hat{p}_{\hat{k}} \end{pmatrix}$$

Note that also the number of subpopulations k is estimated. This book discusses situations where the approach outlined above appears appropriate and develops the suitable theoretical and algorithmic tools for handling mixture models.

Many applications are of the following type: under standard assumptions the population is homogeneous, leading to a simple, one-parametric and *natural* density. Examples include the binomial, the Poisson, the geometric, the exponential, and the normal distributions (with additional variance parameter). If these standard assumptions are violated because of population heterogeneity, mixture models can capture these additional complexities easily. Therefore, C.A.MAN offers most of the conventional densities such as *normal* (common and known different variances), *Poisson, Poisson for SMR models, Binomial, Binomial for rate data, Geometric,* and *Exponential,* among others. To demonstrate these ideas we start with a simple example which has recently found its way into the textbook *Advanced Methods of Marketing Research* (Bagozzi 1995).

Example 1.1 (An introductory example): Data are from a recent branded "hardcandy" new product and concept test, leading to a variable of interest X = "# individual packs of hard candy purchased within the past seven days". Table 1.1 shows its distribution.

# of packages	0	1	2	3	4	5	6	7	8	9	
frequency	102	54	49	62	44	25	26	15	15	10	

# of packages	10	11	12	13	14	15	16	17	18	19	20
frequency	10	10	10	3	3	5	5	4	1	2	1

Table 1.1. Distribution of sold packages

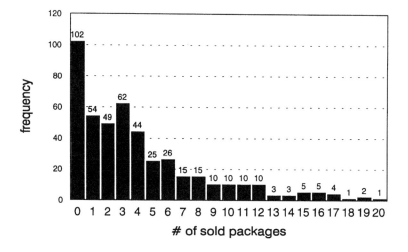

Figure 1.3. Distribution of sold packages.

Typically, the assumption of a Poisson distribution is done for count data under *homogeneity* conditions, e.g., f(x, λ) = Po(x, λ) = e⁻λλˣ/x!. The heterogeneity analysis provided by C.A.MAN delivers a five component mixture distribution, as shown in Figure 1.4.

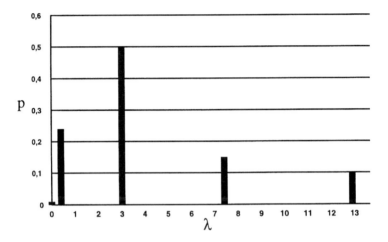

Figure 1.4. NPMLE of mixing distribution for candy data.

These components can easily be interpreted. The two low components correspond to stores with no or almost no sale of the new product. These make about 30% of all stores. There are about 50% with a sale of 3 packages, 15% with about 7.5 packages, and 10% with the large number of 13 packages.

We may also say that there exists a *latent variable,* which might be called PRODUCT SELLING ABILITY of the stores, that explains the extra-Poisson variability in the observed data. Had we known the value of this variable, a stratified analysis would have been possible, leading to homogeneous Poisson models in each stratum. Thus, one can say that the ignorance of an important covariate leads to a mixture of Poisson distributions.

In addition, we see in Figure 1.5 that the mixture model leads to a highly improved fit in comparison to the homogeneous Poisson model.

We summarize the view that mixture models express the presence of extra-population heterogeneity in the following theorem.

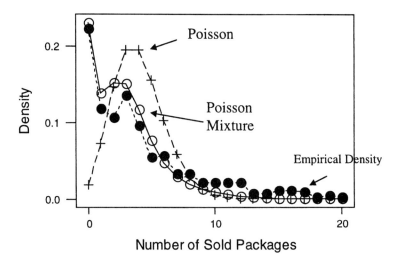

Figure 1.5. Empirical density (solid circle), estimated Poisson density (+), and estimated nonparametric Poisson mixture density (circle)

Theorem 1.1: Let X be an observable variable with density f(x). Let, in addition, Z be an unobservable variable describing the population heterogeneity, e.g., if $Z = z$ then the observation belongs to the zth subpopulation with conditional density f($x \mid Z = z$) = f(x, λ_z). Then

$$f(x) = \int_{-\infty}^{+\infty} f(x, \lambda) P(d\lambda).$$

Proof. The joint density f(x, z) can be written as f(x, z) = f($x \mid z$)f(z) = f(x, λ_z)p_z, where f($x \mid z$) = f(x, λ_z) is the density conditional on membership of subpopulation z. Therefore, the unconditional density f(x) is the *marginal density* $\int_{-\infty}^{+\infty} f(x \mid z)f(z)dz = \int_{-\infty}^{+\infty} f(x, \lambda)P(d\lambda)$, where it is assumed that the probability distribution P has density f(z).

1.2   Some examples

In this section we give some examples of how mixtures can arise and how their appearance can be visualized.

Example 1.2: In a cohort study on 602 pre-school children in NE Thailand it was recorded how often the children showed symptoms of fever, fever and cough, running nose, etc. within a 3 year period (Schelp *et al.* 1986). The data set including its modeling has been discussed on various occasions (Böhning *et al.* 1992, Eilers 1995).

In this case, the nonparametric maximum likelihood estimate turns out to consist of 5 components. They are described in Table 1.2. Note that these 5 components can be easily interpreted as the 5 possible categories of a *latent* variable, HEALTH STATUS say. For example, we see that 48% of the population is estimated to have a "normal" HEALTH STATUS. In addition, since we have explained the extra-Poisson variation, it can be said that conditional on these categories, the count variable NUMBER OF SYMPTOMS follows a Poisson distribution.

j	λ_j	p_j	Interpretation
1	0	0.03	always healthy
2	0.18	0.17	almost always healthy
3	2.82	0.48	normal
4	8.20	0.28	above normal
5	10.15	0.05	high risk for infection

Table 1.2. Estimate of P for preschool children data from NE Thailand

In Figure 1.6 it can be seen that the mixture density provides a highly improved fit in comparison to the homogeneous Poisson density. Indeed, it is becoming quite close to the empirical density. Therefore, mixture densities can also be viewed as alternative ways of finding a *nonparametric estimate of the density*, although this is not the main stream of mixture model development.

In the following we look at some mixture densities for continuous data. Here, the normal distribution is often used. Figure 1.7 shows some normal mixture densities. One point should be noted. Often it is misunderstood that the occurrence of a mixture can be detected by looking for several modes in the empirical distribution. This is not necessarily so, as the third case, namely, 0.5 N(0, 1) + 0.5 N(2, 1), shows. Here the distribution shows just one mode and is symmetric and could be easily (and mistakenly) estimated as a homogeneous normal (with an increased variance).

Mixture densities with one large weight on one component (Figure 1.8) stir up another intense debate. Often these mixtures are difficult to separate from distributions which are by their generating mecha-

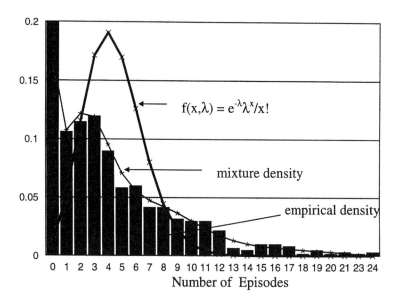

Figure 1.6. Empirical density, estimated Poisson density, and estimated nonparametric Poisson mixture density for acute respiratory infection counts on preschool children from NE Thailand.

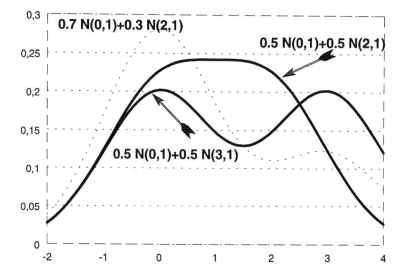

Figure 1.7. Three examples of normal mixtures with two components.

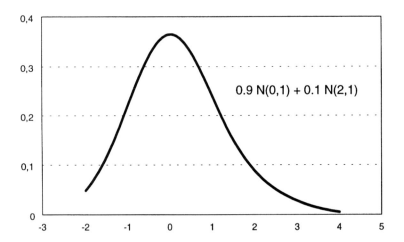

Figure 1.8. Example of normal mixture with large weight on one component

nism asymmetric in their very nature. Take as an example an anthropometric variable such as the body mass index, defined as WEIGHT/(HEIGHT × HEIGHT). This variable is often observed to be skewed. Fitting a mixture model leading to one or more risk groups often provides acceptable model fits and can hardly be distinguished from asymmetric distributional models such as the log-normal. Following the results of the mixture analysis can result in quite erroneous conclusions.

Finally, we look at mixtures of Poisson densities. Figure 1.9 shows 2 examples of Poisson mixtures. One is clearly detectable to be a mixture, since it is bimodal. The other is less clearly a Poisson mixture and other techniques have to be used for diagnosis (such as overdispersion tests).

We consider a data set which has been discussed at various occasions in the literature (Hasselblad 1969, Titterington, Smith, and Makov 1985). It consists of the number of death notices for women aged 80 years and over, from the *Times* newspaper for each day in the three-year period 1910–1912.

Number of notices x_i	0	1	2	3	4	5	6	7	8	9
Frequency n_i	162	267	271	185	111	61	27	8	3	1

Table 1.3. Number of death notices in the *Times* newspaper, 1910–1912

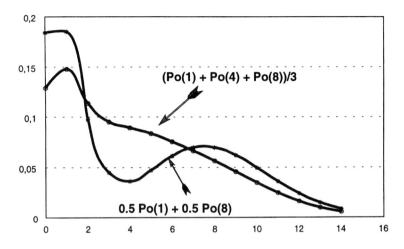

Figure 1.9. Example of two Poisson mixture densities.

Since Hasselblad's analysis it is assumed that a 2-component mixture of Poisson is the appropriate model in this case. Titterington *et al.* (1985) comment that the mixture of two Poisson densities fits quite well with $\hat{P} = \begin{pmatrix} \lambda_1 \lambda_2 \\ p_2 p_2 \end{pmatrix} = \begin{pmatrix} 1.2561 & 2.6634 \\ 0.3599 & 0.6401 \end{pmatrix}$. This is indeed correct; however, it should be pointed out that the nonparametric maximum likelihood estimator has an additional third component, placing some mass at zero:

$$\hat{P} = \begin{pmatrix} 0. & 1.3562 & 2.6984 \\ 0.0068 & 0.3898 & 0.6034 \end{pmatrix}.$$

However, the difference in the log-likelihood is negligible. The real-life interpretation of this mixture model is that there could be different patterns of death in winter and summer, with an additional component for days with no death notices.

1.3   Parametric or nonparametric mixture models?

Consider again the situation which led to the marginal density (1.1). The joint density $f(x, z) = f(x \mid z)p(z) = f(x;\lambda)\, p(\lambda)$ of the observed variate x and the unknown covariate z (describing the population heterogene-

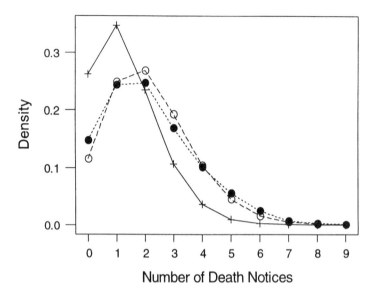

Figure 1.10. Empirical density (solid circle), estimated Poisson density (+), and estimated nonparametric Poisson mixture density (circle).

ity) is expressed as the conditional density $f(x|\lambda) = f(x;\lambda)$ and the density $p(\lambda)$ describing the distribution of the parameter λ in the population. Suppose that $p(\lambda)$ is modeled as a parametric, continuous density with some mean parameter μ and variance τ^2. The representatives of this approach usually give a variety of reasons for doing so including the following two:

(a) In many applications it makes no sense to assume that the population heterogeneity is represented by some discrete quantity having a discrete distribution. It is more appropriate to assume a continuous quantity with associated continuous distribution. Since it can be shown that the nonparametric estimator of the mixing distribution is necessarily discrete (Lindsay 1983), it appears reasonable to proceed parametrically.

(b) In some specific applications the parametric approach leads to marginal densities which are mathematically simple to treat.

In analogy to (1.1) the marginal density in the parametric case is

$$f(x, \Phi) = \int_{-\infty}^{+\infty} f(x;\lambda)p(\lambda)d\lambda \tag{1.2}$$

where Φ is the vector of parameters which is involved in $p(\lambda)$. We demonstrate the procedure with one example.

Example 1.3: (Poisson-gamma) Let $f(x;\lambda) = Po(x, \lambda)$ and $p(\lambda) = p(\lambda;\theta, \kappa)$ $= \theta^{-\kappa}\lambda^{\kappa-1}e^{-\lambda/\theta}/\Gamma(\kappa)$. Note that $\Phi = (\theta, \kappa)^T$ and $\mu = \kappa\theta$, $\tau^2 = \kappa\theta^2 = \mu\theta = \mu^2/\kappa$. It follows that

$$f(x, \Phi) = \int_0^{+\infty} f(x;\lambda)p(\lambda)d\lambda = \int_0^{+\infty} e^{-\lambda}\lambda^x/x!\; \theta^{-\kappa}\lambda^{k-1}e^{-\lambda/\theta}/\Gamma(\kappa)d\lambda$$

$$= \frac{\theta^{-\kappa}}{\Gamma(x+1)\Gamma(\kappa)}\int_0^{+\infty} e^{-\lambda(1+1/\theta)}\lambda^{x+\kappa-1}d\lambda$$

$$= \frac{\theta^{-\kappa}}{\Gamma(x+1)\Gamma(\kappa)}\int_0^{+\infty} e^{-z}z^{x+\kappa-1}(\theta/(1+\theta))^{x+\kappa-1}d\lambda,$$

(with $z = \lambda(\theta + 1)/\theta$)

$$= \frac{\theta^{-\kappa}}{\Gamma(x+1)\Gamma(\kappa)}\int_0^{+\infty} e^{-z}z^{x+\kappa-1}(\theta/(1+\theta))^{x+\kappa-1}\theta/(1+\theta)dz,$$

(because $dz/d\lambda = (\theta + 1)/\theta$)

$$= \frac{(\theta+1)^{\kappa}\Gamma(x+\kappa)}{\Gamma(x+1)\Gamma(\kappa)}(\theta/(1+\theta))^x,$$

because $\int_0^{+\infty} e^{-z}z^{x+\kappa-1}dz = \Gamma(x+\kappa)$ by definition of the gamma function. The latter expression can be written as

$$f(x, \Phi) = \frac{\Gamma(x+\kappa)}{\Gamma(x+1)\Gamma(\kappa)}(1+\mu/\kappa)^{-\kappa}(\mu/(\mu+\kappa))^x$$

$$= \frac{\Gamma(x+\kappa)}{\Gamma(x+1)\Gamma(\kappa)}p^{\kappa}(1-p)^x$$

the *negative binomial* distribution. The negative binomial distribution is a well-known distribution. In particular, the mean $E(x) = \mu$ and $Var(x)$ $= \mu + \mu^2/\kappa$ are known. From here it is easy to derive moment estimates for μ and κ: $\hat{\mu}_{MO} = \bar{x}$, $\hat{\kappa}_{MO} = \bar{x}^2/(S^2 - \bar{x})$, where $\bar{x}$ and S^2 are sample mean and sample variance.

Example 1.4: (Aflatoxin in peanuts) Giesbrecht and Whitaker (1998) study a variety of distributions for the Aflatoxin-level (AT) in or on peanut kernels. In peanut production and marketing a frequently occurring problem is the Aflatoxin-producing mold *Aspergillus flavus*, which can lead to diverse toxic effects in human beings when consumed uncontrolled. A common problem is to determine the appropriate distributional model for the AT-level. This is of interest for a variety of reasons including sample size determination or fitting covariate models. Consider the frequency table below (Table 1.4) which is from a larger collection of lots (data according to Giesbrecht and Whitaker 1998).

AT-level $\times\ 10^{-1}$	0	1	2	3	4	5	6	8	9	10	11	15
frequency	8	6	5	3	2	3	1	3	1	4	2	1

AT-level $\times\ 10^{-1}$	19	20	22	28	29	36	40	53	58	82	97
frequency	1	1	1	1	1	1	1	1	1	1	1

Table 1.4. Sample of size 50 of AT-levels in or on peanut kernels (first row AT-level $\times\ 10^{-1}$, second row frequency)

As can be seen in Figure 1.11 the Poisson distribution does not fit well, which is also indicated by the first two moments $\bar{x} = 12.96$ and $S^2 = 420.2$ for which we have strong *overdispersion* ($S^2 > \bar{x}$). In fact, the overdispersion test $O_T = \sqrt{(n-1)/2}(S^2 - \bar{x})/\bar{x} = 155.4$ is highly significant. The fit improves considerably if the negative binomial is considered. Figure 1.12 shows the result of the fit.

Let $F(x, \hat{P}) = \int_{-\infty}^{x} f(y;\hat{P})dy$ denote the distribution function of x with estimated mixture density $f(x, \hat{P})$ whereas

$$F(x, \hat{\Phi}) = \int_{-\infty}^{x} \frac{\Gamma(y + \hat{\kappa})}{\Gamma(y + 1)\Gamma(\hat{\kappa})} \hat{p}^{\hat{\kappa}}(1 - \hat{p})^{y}dy$$

is the distribution function of the negative binomial distribution ($\hat{\Phi}$ are the moment estimators $(\hat{\mu}, \hat{\kappa})^{T} = (\bar{x}, \bar{x}^2/(S^2 - \bar{x}))^{T}$).

Consider the residual $\varepsilon(x) = \hat{F}(x) - F(x, \hat{\Phi})$ in the case of the negative binomial, or $\varepsilon(x) = \hat{F}(x) - F(x, \hat{P})$ in the nonparametric case. It is quite visible that there is further improvement in the AT-levels from 10 to 20.

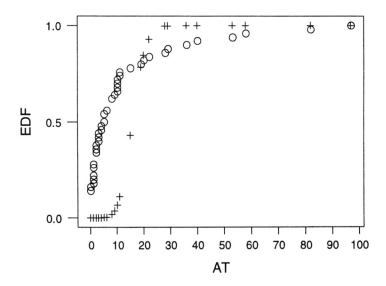

Figure 1.11. Empirical distribution function (EDF) and Poisson distribution function for the AT-levels of Table 1.4 (open circle is EDF).

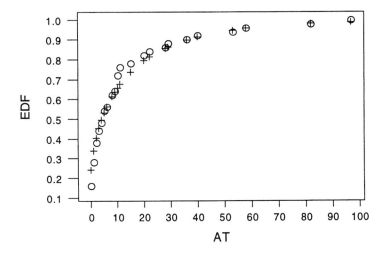

Figure 1.12. Empirical distribution function (EDF) and negative binomial distribution function for the AT-Levels of Table 1.4 (open circle is EDF).

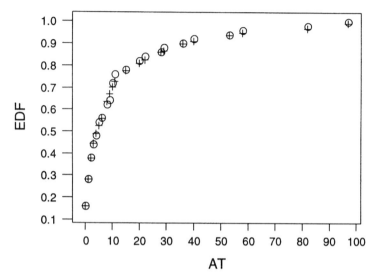

Figure 1.13. Empirical distribution function (EDF) and distribution function of non-parametrically mixed Poisson for the AT-levels of Table 1.4 (open circle is EDF).

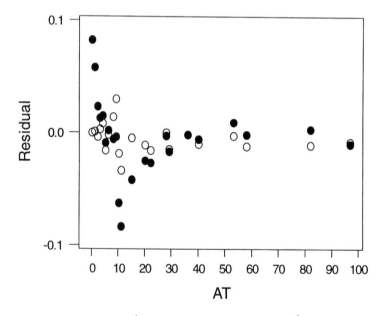

Figure 1.14. Residual (F(AT, $\hat{P}$)-EDF(AT) is open circle, F(AT, $\hat{\Phi}$) – EDF(AT) is solid circle) for the AT-Levels of Table 1.4.

Example 1.5: (Beta-binomial) Let $f(x; \lambda) = \binom{n}{x} \lambda^x (1 - \lambda)^{n-x}$ be the binomial distribution and $p(\lambda) = \Gamma(\alpha + \beta)/(\Gamma(\alpha)\Gamma(\beta))\lambda^{\alpha - 1}(1 - \lambda)^{\beta - 1}$ the beta density, where $\Gamma(t) = \int_0^\infty y^{t-1}e^{-y}dy$ is the conventional Gamma function. Then

$$f(x, \Phi) = \int_{-\infty}^{+\infty} f(x;\lambda)p(\lambda)d\lambda$$

$$= \int_{-\infty}^{+\infty} \binom{n}{x}\lambda^x(1 - \lambda)^{n-x}\frac{\Gamma(\alpha + \beta)}{\Gamma(\alpha)\Gamma(\beta)}\lambda^{\alpha - 1}(1 - \lambda)^{\beta - 1}d\lambda$$

$$= \binom{n}{x}\frac{\Gamma(\alpha + \beta)}{\Gamma(\alpha)\Gamma(\beta)} \int_{-\infty}^{+\infty} \lambda^{x+\alpha - 1}(1 - \lambda)^{n-x+\beta - 1}d\lambda$$

$$= \binom{n}{x}\frac{\Gamma(\alpha + \beta)}{\Gamma(\alpha)\Gamma(\beta)} \int_{-\infty}^{+\infty} \frac{\Gamma(x + \alpha + n - x + \beta)}{\Gamma(x + \alpha)\Gamma(n - x + \beta)}\lambda^{x+\alpha - 1}(1 - \lambda)^{n-x+\beta - 1}d\lambda$$

$$\times \frac{\Gamma(x + \alpha)\Gamma(n - x + \beta)}{\Gamma(x + \alpha + n - x + \beta)} = \binom{n}{x}\frac{\Gamma(\alpha + \beta)\Gamma(x + \alpha)\Gamma(n - x + \beta)}{\Gamma(\alpha)\Gamma(\beta)\Gamma(x + \alpha + n - x + \beta)},$$

since the integral is over the beta distribution with parameters $(x + \alpha)$ and $(n - x + \beta)$.

$$f(x, \Phi) = f(x, \alpha, \beta) = \binom{n}{x}\frac{\Gamma(\alpha + \beta)\Gamma(x + \alpha)\Gamma(n - x + \beta)}{\Gamma(\alpha)\Gamma(\beta)\Gamma(x + \alpha + n - x + \beta)}$$

is a special distribution called *beta-binomial* with mean $E(x) = \mu = \alpha/(\alpha + \beta)$ and variance $\text{Var}(x) = n\alpha\beta/(\alpha + \beta)^2 (\alpha + \beta + n)/(\alpha + \beta + 1) = n\mu(1 - \mu) (1 + \theta n)/(1 + \theta)$, where $\theta = 1/(\alpha + \beta)$. There is another representation of $\text{Var}(x)$ which involves the *intraclass correlation* coefficient $\rho = \theta/(1 + \theta)$, namely, $\text{Var}(x) = n\mu(1 - \mu) (1 + \rho(n - 1))$. Specifically, this latter expression demonstrates that the beta-binomial is an *overdispersion* model ($\rho \geq 0$).

Estimates for α and β can readily be provided using moment estimators. Since $E(x) = \mu$, we find that $\hat{\mu}_{MO} = (1/N)\sum_{i=1}^{N} x_i/n$ for a sample of binomial distributions $x_1, \ldots, x_N$, each being of size n.

Since $\text{Var}(x/n) = \mu(1 - \mu)/n (1 + \rho(n - 1))$, we find

$$\hat{\rho}_{MO} = S^2/[(n-1)\hat{\mu}_{MO}(1 - \hat{\mu}_{MO})/n] - 1/(n - 1) \tag{1.3}$$

With the transformation $\theta = \rho/(1 - \rho)$ we find

$$\hat{\theta}_{MO} = \frac{S^2 - \hat{\mu}_{MO}(1 - \hat{\mu}_{MO})/n}{\hat{\mu}_{MO}(1 - \hat{\mu}_{MO}) - S^2}. \qquad (1.4)$$

See also Carlin and Louis (1996, Chapter 3.3.2).

1.4   Connection to empirical Bayes estimation

Consider Bayes' theorem relating conditional density and prior distribution with the posterior:

$$f(\lambda|x) = \frac{f(x|\lambda)p(\lambda)}{f(x;P)} = \frac{f(x|\lambda)p(\lambda)}{\displaystyle\int_{-\infty}^{+\infty} f(x;\lambda)p(\lambda)d(\lambda)} \qquad (1.5)$$

if P is a parametric distribution with density $p(\lambda)$, or

$$f(\lambda_j|x) = \frac{f(x|\lambda_i)p_i}{f(x;P)} = \frac{f(x|\lambda_i)p_i}{\displaystyle\sum_{l=1}^{k} f(x;\lambda_l)p_l} \qquad (1.0)$$

if P is nonparametric and discrete. In classical Bayesian inference the prior $p(\lambda)$ is considered to express the subjective opinion of the Bayesian statistician. However, $p(\lambda)$ can be also thought of as the distribution reflecting the distribution of a diversity of parameters in the population, in other words, expressing the heterogeneity of the population with respect to λ. If this is so, then the heterogeneity distribution of λ, namely, the distribution P in the normalizing constant $f(x, P)$ in Bayes' theorem, can be estimated, for example, by maximum likelihood. This is precisely the *empirical* Bayes approach.

Having $f(\lambda|x)$ or its estimate available, various statistical measures can be computed, for example the posterior mean, also called *the empirical Bayes* estimate of λ:

$$x^{EB} = E(\lambda|x) = \int_{-\infty}^{+\infty} \lambda f(\lambda|x)d\lambda = \int_{-\infty}^{+\infty} \lambda \frac{f(x|\lambda)p(\lambda)}{\displaystyle\int_{-\infty}^{+\infty} f(x;\lambda)p(\lambda)d\lambda} d\lambda \qquad (1.7)$$

if $p(\lambda)$ is a continuous parametric density and

$$x^{EB} = E(\lambda|x) = \sum_{j=1}^{k} \lambda_j \frac{f(x|\lambda_j)p_j}{f(x;P)}$$ (1.8)

if P is a discrete probability distribution. As an example, let us consider the Poisson density: $f(x, \lambda) = e^{-\lambda} \lambda^x/x!$. Then, $x^{EB} = (x + 1) f(x + 1, P)/f(x, P)$. If P is left nonparametric, this is about all that can be said.* If parametric assumptions are met for P, then x^{EB} takes on specific forms.

Theorem 1.2: Let $p(\lambda)$ be a Γ-distribution with parameters μ and κ:

$$p(\lambda) = \lambda \kappa^{-1} e^{-\lambda/\theta} \theta^{-\kappa}/\Gamma(\kappa) \text{ with } \theta = \mu/\kappa.$$

If $f(x, \lambda)$ is the Poisson, then $x^{EB} = (x + \kappa)/(1 + 1/\theta) = (x + \kappa)/(1 + \kappa/\mu)$.

Proof. According to example 1.3 we have that

$$f(x, P) = \int_{-\infty}^{+\infty} f(x;\lambda)p(\lambda)d\lambda = \frac{\Gamma(x+\kappa)}{\Gamma(x+1)\Gamma(\kappa)} p^{\kappa}(1-p)^x,$$

with $p = \kappa/(\kappa + \mu)$. Now,

$$x^{EB} = (x+1)f(x+1, P)/f(x, P)$$

$$= (x+1)\frac{\Gamma(x+1+\kappa)}{\Gamma(x+2)\Gamma(\kappa)} p^{\kappa}(1-p)^{x+1} / \left\{ \frac{\Gamma(x+\kappa)}{\Gamma(x+1)\Gamma(\kappa)} p^{\kappa}(1-p)^x \right\}$$

$$= (x+1)(x+\kappa)/(x+1)(1-p)\frac{\Gamma(x+\kappa)}{\Gamma(x+1)\Gamma(\kappa)} p^{\kappa}(1-p)^x /$$

$$\left\{ \frac{\Gamma(x+\kappa)}{\Gamma(x+1)\Gamma(\kappa)} p^{\kappa}(1-p)^x \right\} = (x+\kappa)\mu/(\kappa+\mu).$$

---

* Note, however, that it is possible to use the relative frequency density # (xs equal to x)/n to estimate $f(x, P)$. This leads to $x^{EB} = (x + 1)$ #(xs equal to $x + 1$)/#(xs equal to x) which was proposed by Robbins (1955). Carlin and Louis (1996) comment on this estimator that it illustrates a model for which the analyst can make empirical Bayes inferences *with no knowledge* of P whatsoever.

This completes the proof.

As a consequence we note that $\text{Var}(x^{EB}) = \text{Var}(x)/(1 + 1/\theta)^2$, e.g., we have a smaller variance for x^{EB} in comparison to our original datum. The *shrinkage* factor $1/(1 + 1/\theta)^2$ is the greater, the more *homogeneous* the population is. If the variance of the Γ-distribution is 0 (e.g., $\mu^2/\kappa = 0$, or κ infinite given μ finite), then $x^{EB} = (x + \kappa)\mu/(\kappa + \mu) = (x/\kappa + 1)/(\mu/\kappa + 1) \mu \to \mu$ for $\kappa \to \infty$. Conversely, if κ is small, implying the presence of parameter *heterogeneity*, then $x^{EB} = (x + \kappa)\mu/(\kappa + \mu) \to x$ if $\kappa \to 0$. This has the conventional interpretation that in the case of strong heterogeneity, not much information about x can be gained from the other sample members.

Theorem 1.3: Let P be a distribution on λ with mean μ and variance τ^2. Also, $f(x, \lambda) = f(x \mid \lambda)$ be a parametric density for random variate x. Then

(a) $\text{Var}(x^{EB}) \leq \text{Var}(x)$ if x^{EB} is given by (1.7) or (1.8).

(b) If $\tau^2 = 0$, $x^{EB} = \mu$.

Proof. Part (b) follows from the fact that P is a one-point distribution at μ if $\tau^2 = 0$.

To show a) we first denote with Λ the random variable associated with the distribution of the parameter λ in the population. We have by definition of the variance:

$$\text{Var}(\Lambda \mid x) = E(\Lambda^2 \mid x) - E(\Lambda \mid x)^2$$

It follows that $E_x [\text{Var}(\Lambda \mid x)] = E_x [E(\Lambda^2 \mid x)] - E_x [E(\Lambda \mid x)^2]$, where E_x is the expected value with respect to $f(x, P)$. Now,

$$E_x [E(\Lambda^2 \mid x)] = E(\Lambda^2) = \text{Var}(\Lambda) + E(\Lambda)^2 = \tau^2 + \mu^2,$$

leading to

$$E_x [\text{Var}(\Lambda \mid x)] = \text{Var}(\Lambda) + E(\Lambda)^2 - E_x [E(\Lambda \mid x)^2]$$

$$= \tau^2 - (E_x [E(\Lambda \mid x)^2] - E(\Lambda)^2)$$

$$= \tau^2 - (E_x [E(\Lambda \mid x)^2] - [E_x E(\Lambda \mid x)]^2)$$

$$= \tau^2 - \text{Var}_x E(\Lambda \mid x)$$

Therefore, Var (x^{EB}) = Var$_x$ E$(\Lambda \mid x)$ = τ^2 – E$_x$ [Var$(\Lambda \mid x)$] $\leq \tau^2$.

A direct argument shows that Var(x) = E [Var$(x \mid \lambda)$] + $\tau^2 \geq \tau^2$, where E [Var$(x \mid \lambda)$] = $\int$Var$(x \mid \lambda) P(d\lambda)$ and $\tau^2 = \int (\lambda - \mu)^2 P(d\lambda)$. This completes the proof.

Usually, it will be necessary to replace the theoretical parameters involved in the prior distribution by their sample estimates. This can be accomplished by finding the mixture maximum likelihood estimate $\hat{P}$. This leads to

$$\hat{x}^{EB} = \int_{-\infty}^{+\infty} \lambda \frac{f(x;\lambda)\hat{p}(\lambda)}{\int_{-\infty}^{+\infty} f(x;\lambda)\hat{p}(\lambda)d\lambda} d\lambda$$

or

$$\hat{x}^{EB} = \sum_{j=1}^{k} \hat{\lambda}_j \frac{f(x \mid \hat{\lambda}_j)\hat{p}_j}{f(x;\hat{P})},$$

in case we are having the discrete, nonparametric maximum likelihood estimator as an estimator of the prior distribution (see also Laird 1982).

According to Theorem 1.3, we can expect more differences between the empirical distribution and the distribution of the empirical Bayes estimates if there is less heterogeneity in the parameter distribution of the λs, in the extreme case of homogeneity in which the distribution of the empirical Bayes estimators reduces to a one point distribution, as shown in Figure 1.16 for mortality data of Berlin in the year 1988 (with ICD430-438 (stroke) as cause of death).

1.5   Classification using posterior Bayes

Having found the mixing distribution P and its estimate $\hat{P}$, it is not clear to which component of the mixing distribution each observation x belongs, and one might be interested in classifying each observation x into one of the components of the mixing distribution. This is of particular interest if P is discrete, nonparametric, since in this case

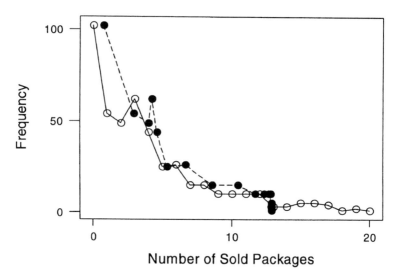

Figure 1.15. Distribution of sold packages (circle) and corresponding empirical Bayes estimates of sold packages (solid circle) (data from Example 1.1).

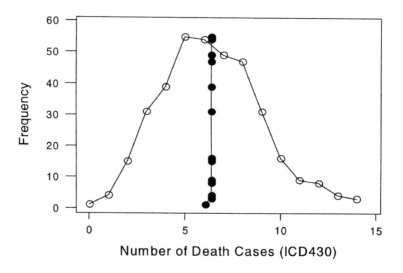

Figure 1.16. Distribution of number of death cases per day in Berlin 1988 with ICD430-438 as cause of death (circle) and corresponding empirical Bayes estimates (solid circle).

the mixture components can be thought of *disjoint classes* into which the population can be partitioned. Classification can be accomplished by means of the posterior distribution

$$f(\lambda_j|x) = \frac{f(x|\lambda_j)p_j}{f(x;P)} = \frac{f(x|\lambda_j)p_j}{\sum_{l=1}^{k} f(x;\lambda_l)p_l} \qquad (1.9)$$

such that x is classified into component j with

$$f(\lambda_j | x) = \max_i f(\lambda_i | x) \qquad (1.10)$$

with $1 \le i \le k$. Again, to put this into practical terms we have to replace P by its maximum likelihood estimator $\hat{P}$ and (1.10) takes the form

$$f(\hat{\lambda}_j|x) = \max_i f(\hat{\lambda}_i|x) = \max_i \frac{f(x|\hat{\lambda}_i)\hat{p}_i}{f(x;\hat{P})} . \qquad (1.11)$$

Example 1.6: A study evaluated the effectiveness of Avon's Skin-So-Soft® (SSS) as a flea repellent for dogs (Fehrer and Halliwell 1987). Table 1.5 shows the flea counts for 10 dogs one day after a sponge dip with SSS or water.

Dog	1	2	3	4	5	6	7	8	9	10
Flea Count	88	30	26	49	106	109	13	121	142	147
Dip	SSS	SSS	SSS	SSS	SSS	W	W	W	W	W
Classification	1	1	1	1	2	2	1	2	2	2

Table 1.5. Flea counts for 10 dogs 1 day after a sponge dip with SSS or water

Although the difference in the means is not significant, it is of a size which can be suspected to become significant at a more appropriate sample size (in other words, the test lacks power in this case). Now, let us suppose that we were *not* able to say which flea count belongs to which of the two groups (for example, it might have happened that the treatment allocation had been lost). To what degree is the mixture approach able to reconstruct the assignment to groups? If one estimates the mixture distribution non-parametrically with $f(x, \lambda) = Po(x, \lambda)$ as mixture kernel, then a 2-component solution is found. Classification using the maximum posterior allocation rule (1.11) leads to row four of Table 1.5. All reclassifications are *correct*, other than the flea count for dog 7 which is wrongly assigned to the SSS group and for dog 5 which is wrongly assigned to the Water group.

The foregoing example demonstrates that ignorance of a relevant factor leads to *heterogeneity*. Thus, conversely, the occurrence of heterogeneity can be taken as indicative of the presence of an *unobserved* factor.*

1.6   Analysis of geographical heterogeneity in live-birth ratio in Thailand

In this section we are interested in applying the foregoing ideas to a more complete example. Live-birth (male-female) ratios is a standard measure used in demography. Live-birth ratios have been recently considered more frequently as a potential indicator for environmental hazards of various nature (Williams, Lawson, and Lloyd 1992). Not only does population growth depend on it, the live-birth ratio is associated with a number of factors including parental occupation, exposure to smoking and drinking, radiation, war and post-war conditions as well as social class (Levin 1987). Although male births tend to exceed female births in human societies, this ratio is neither uniform nor static. The ratio varies between countries, e.g., 1.04 for black Americans, 1.06 in Scotland, and 1.15 in Korea. For Thailand it is 1.05. Also, the ratio varies in time, e.g., it is usually higher in war and post-war times (Williams, Lawson, and Lloyd 1992, Levin 1987). The mechanisms which cause the sex ratio to vary have not yet been clarified, but there may be at least three factors, possibly interconnected, which are described by Williams *et al.* (1992) as: "The importance of *hormonal influences* has been indicated, for example, by an excess of male births associated with high levels of parental testosterone, maternal oestrogen and paternal gonadotrophin; by contrast, high levels of maternal gonadotrophin are associated with an excess of female births. Other possible influences, which might act through the hormonal route, are *maternal diet* and *exposure to chemicals* such as the pesticide DBCP. Abnormal sex ratios have been clearly associated with some *occupational environments*. An excess of male births is reported for taxation experts, chartered accountants, environmental health officers, and certain categories of engineer; whereas an excess of female births is associated with messengers, librarians, quantity surveyors and with tertiary education teachers. It is plausible that abnormal sex ratios could occur also in residents in areas at risk from airborne pollutants

---

*This statement has to be taken with care, however. There might be other reasons for the occurrence of heterogeneity, such as that the distributional model of choice is *not* correct.

from industries.... If sex ratios *(of births, D.B.)* are affected by residential exposure to airborne pollution, the detection of such abnormalities would constitute a simple screening procedure to alert medical and environmental health authorities to insidious hazards to health."

In what follows we are interested in developing a monitoring device for live-birth ratios. The live-birth ratio is defined λ_M/λ_F, where λ_M is the probability of a *male* live-birth, whereas λ_F is the probability of a *female* live-birth.*

If for some geographical area n_M male and n_F female live-births are observed, it is clear that — conditional on the total number of live-births $n = n_M + n_F$ — the number of male live-births $x = n_M$ is binomially distributed with parameter λ_M :

$$f(x, \lambda_M) = \binom{n}{x}\lambda_M^x \lambda_F^{(n-x)} . \tag{1.12}$$

Let for simplicity $\lambda = \lambda_M$. A geographical analysis of live-birth ratios requires data on a small area. Then, if deviations from the national or world live-birth ratio are observed, this might be taken as some evidence for further investigations into various sources such as occupational and environmental hazards. In the following a geographical analysis of live-birth ratios is presented for the Kingdom of Thailand. Table 1.6 presents data on the smallest available level, in this case the level of province. There are 73 provinces. The explanations for the 9 columns in Table 1.6 are given in Table 1.7.

Let $\hat{\lambda}_{iM} = n_{M,i}/n_i$ the estimated proportion of male live-births in area i. As a first diagnostic measure the values of the Z-statistics might be used which is defined as

$$Z_i = \frac{\hat{\lambda}_{iM} - \lambda}{\sqrt{\lambda(1-\lambda)/n_i}} \quad \text{for } i = 1, ..., N. \tag{1.13}$$

As a value for λ, $\lambda = 0.513$ is used, which corresponds to the national standard of Thailand (the sex ratio is $\lambda/(1 - \lambda) = 1.05$ which is well within the range of world-wide sex ratios). Under homogeneity the Z-values should follow a standard normal distribution, at least approximately. In Table 1.6 we find a number of values which are below or above the 5%-critical values of 1.96. However, even under homogeneity we can expect a number of Z-values (about 3–4) which fall into the

---

*The ratio λ_M/λ_F is often given in the form $\lambda_M/\lambda_F \times 100$.

I	II	III	IV	V	VI	VII	VIII	IX
10	39333	37266	0.5135	0.2271	2	−3.7467	95	1
11	5672	5150	0.5241	2.2967	3	−3.4303	57	1
12	3947	3870	0.5049	−1.4428	2	−3.1486	86	1
13	2944	2727	0.5191	0.9115	2	−2.4622	52	1
14	5082	4702	0.5194	1.2540	2	−2.4269	44	1
15	1881	1772	0.5149	0.2221	2	−1.7041	61	2
16	5408	5050	0.5171	0.8253	2	−1.6477	31	2
17	1559	1472	0.5143	0.1398	2	−1.6023	49	2
18	2708	2423	0.5277	2.1052	2	−1.5533	62	2
19	4214	3989	0.5137	0.1145	2	−1.4428	12	2
20	5397	4986	0.5197	1.3678	2	−1.4091	53	2
21	3609	3204	0.5297	2.7479	3	−1.3209	42	2
22	3164	2864	0.5248	1.8331	2	−1.3054	41	2
23	1191	1090	0.5221	0.8654	2	−1.2925	90	2
24	4774	4500	0.5147	0.3256	2	−1.2220	25	2
25	6995	6778	0.5078	−1.2220	2	−1.1773	83	2
26	1709	1622	0.5130	−0.0026	2	−1.1402	48	2
30	22494	21522	0.5110	−0.8566	2	−0.9266	77	2
31	13312	12893	0.5079	−1.6477	2	−0.9139	55	2
32	12649	11946	0.5142	0.3794	2	−0.8566	30	2
33	13364	12623	0.5142	0.3789	2	−0.7704	71	2
34	19984	18866	0.5143	0.5151	2	−0.6148	96	2
35	4930	4681	0.5129	−0.0251	2	−0.5257	43	2
36	9084	8490	0.5169	1.0125	2	−0.5173	75	2
40	15229	14416	0.5137	0.2170	2	−0.5071	50	2
41	17867	17194	0.5095	−1.3054	2	−0.4693	51	2
42	4971	4845	0.5064	−1.3209	2	−0.3602	63	2
43	7875	7537	0.5109	−0.5257	2	−0.2596	92	2
44	7662	7563	0.5032	−2.4269	1	−0.2372	93	2
45	10584	10012	0.5138	0.2308	2	−0.1136	58	2
46	8067	7557	0.5163	0.8099	2	−0.1126	94	2
47	8694	8160	0.5158	0.7167	2	−0.0289	85	2
48	5627	5457	0.5076	−1.1402	2	−0.0251	35	2
49	2351	2338	0.5013	−1.6023	2	−0.0026	26	2
50	10003	9562	0.5112	−0.5071	2	0.0565	66	2
51	3046	2926	0.5100	−0.4693	2	0.1085	84	2
52	5736	5700	0.5015	−2.4622	1	0.1145	19	2

Table 1.6 Geographical heterogeneity in live-births in Thailand 1990.

I	II	III	IV	V	VI	VII	VIII	IX
53	3438	3376	0.5045	−1.4091	2	0.1173	67	2
54	3776	3405	0.5258	2.1615	3	0.1398	17	2
55	3288	3192	0.5074	−0.9139	2	0.2170	40	2
56	3585	3278	0.5223	1.5387	2	0.2221	15	2
57	7917	7934	0.4994	−3.4303	1	0.2270	10	2
58	1551	1478	0.5120	−0.1136	2	0.2308	45	2
60	8373	7711	0.5205	1.9022	2	0.2449	74	2
61	2272	2268	0.5004	−1.7041	2	0.3064	81	2
62	6102	5957	0.5060	−1.5533	2	0.3256	24	2
63	2869	2749	0.5106	−0.3602	2	0.3275	76	2
64	4593	4224	0.5209	1.4734	2	0.3789	33	2
65	6537	5978	0.5223	2.0705	2	0.3794	32	2
66	4255	4033	0.5133	0.0565	2	0.4087	80	2
67	7668	7263	0.5135	0.1173	2	0.4946	73	2
70	6084	5358	0.5317	3.9897	3	0.5151	34	2
71	5943	5721	0.5095	−0.7704	2	0.7167	47	2
72	6437	6025	0.5165	0.7700	2	0.7700	72	2
73	5082	4775	0.5155	0.4946	2	0.8099	46	2
74	2590	2441	0.5148	0.2449	2	0.8253	16	2
75	1183	1147	0.5077	−0.5173	2	0.8654	23	2
76	3103	2920	0.5151	0.3275	2	0.9115	13	2
77	3611	3503	0.5075	−0.9266	2	0.9659	82	2
80	13239	12500	0.5143	0.4087	2	1.0125	36	2
81	2913	2742	0.5151	0.3064	2	1.2540	14	2
82	2133	1964	0.5206	0.9659	2	1.3678	20	2
83	1473	1460	0.5022	−1.1773	2	1.4388	91	2
84	6493	6150	0.5135	0.1085	2	1.4734	64	2
85	1124	1068	0.5127	−0.0289	2	1.5387	56	2
86	3185	3269	0.4934	−3.1486	1	1.8331	22	2
90	10086	9749	0.5084	−1.2925	2	1.9022	60	2
91	2566	2337	0.5233	1.4388	2	2.0705	65	2
92	5536	5280	0.5118	−0.2596	2	2.1052	18	2
93	4092	3904	0.5117	−0.2372	2	2.1615	54	3
94	5998	5704	0.5125	−0.1126	2	2.2967	11	3
95	3701	3829	0.4915	−3.7467	1	2.7479	21	3
96	5516	5297	0.5101	−0.6148	2	3.9897	70	3

Column	Variable name
I	Province
II	Number of male live-births
III	Number of female live-births
IV	Estimated proportion of male live-births
V	Z-statistic
VI	Classification into Class 1, 2 or 3
VII	Sorted according to Z-statistic
VIII	Province associated with sorted Z-statistic
IX	Classification associated with sorted Z-statistic

Table 1.7. Explanation of the 8 columns in Table 1.5

critical region. What is in question here relates to the overall performance of the homogeneous binomial model. We therefore consider the
question of modeling *heterogeneity* for this situation.

The corresponding model assumes the existence of an (unknown)
number of subpopulations k with parameters λ_j as the probability of a
male live-birth in this subpopulation j and a population weight p_j
representing the proportion with which this subpopulation is being
represented in population, $j = 1, ..., k$.

Let a latent variable Z describe the subpopulation membership.
According to (1.1) the unconditional density f(x) is the *marginal density*

$$f(x, P) = \sum_{j=1}^{k} f(x;\lambda_j)p_j = \sum_{j=1}^{k} \binom{n}{x}\lambda_j^x(1-\lambda_j)^{(n-x)}p_j$$

Consequently, the likelihood is given as

$$\prod_{i=1}^{N}\sum_{j=1}^{k} \binom{n_i}{x_i}\lambda_j^{x_i}(1-\lambda_j)^{n_i-x_i}p_j$$

which needs to be maximized in the number of subpopulations k, as
well as in the parameters $\lambda_1, ..., \lambda_k$, receiving weight $p_1, ..., p_k$, respectively. C.A.MAN provides the maximum likelihood estimator
$\hat{P} = \begin{pmatrix} \hat{\lambda}_1...\hat{\lambda}_k \\ \hat{p}_1...\hat{p}_k \end{pmatrix}$ of this mixing distribution P. Table 1.8 provides the estimate of the mixing distribution. Here, the number of subpopulations
is estimated to be $k = 3$. If one compares the mixture log-likelihood
with the single binomial log-likelihood, a difference of about 11 can be

noted, which, if multiplied by 2, provides a highly significant log-likelihood ratio statistic.

λ_j	p_j	Log-likelihood
0.500026	0.0805	−411.10760
0.512998	0.8371	
0.525192	0.0824	

λ	p	Log-likelihood
0.512885	1.0000	−422.64860

Table 1.8. Estimates of C.A.MAN for the mixing distribution P; for comparison the MLE of λ in the homogeneous case ($k = 1$) is given as well

This result can be therefore taken as evidence that there is some degree of heterogeneity in the live-birth ratios. Whereas about 84% of the provinces show correspondence to the standard value, about 8% fall in the class with decreased proportion of male live-birth, and 8% fall into the class of increased male live-births.

Let us now compute the probability that the ith area belongs to the jth subpopulation, given the observed data. Let Z_{ij} describe the unobserved indicator that area i is in subpopulation j ($Z_{ij} = 1$). According to Section 1.5 and the classification rule (1.11) we find,

$$
\begin{aligned}
\Pr(Z_{ij} = 1 | \mathbf{x}, \mathbf{n}) &= \frac{\binom{n_i}{x_i}\lambda_j^{x_i}(1 - \lambda_j)^{n_i - x_i}\Pr(Z_{ij} = 1)}{\sum_{l=1}^{k}\binom{n_i}{x_i}\lambda_l^{x_i}(1 - \lambda_l)^{n_i - x_i}\Pr(Z_{il} = 1)} \\[2ex]
&= \frac{\binom{n_i}{x_i}\lambda_j^{x_i}(1 - \lambda_j)^{n_i - x_i}p_j}{\sum_{l=1}^{k}\binom{n_i}{x_i}\lambda_l^{x_i}(1 - \lambda_l)^{n_i - x_i}p_l}
\end{aligned}
\tag{1.14}
$$

These posterior probabilities are often denoted as w_{ij}. The classification mechanism (1.14) according to maximum posterior probability, which is frequently employed, categorizes every area into that subpopulation j for which the posterior probability is largest: $w_{ij} = \max_{1 \leq l \leq k} w_{il}$.

This is done for the live-birth ratio data of Thailand and the results are found in column VI of Table 1.6. Five areas are classified into the group of reduced birth-proportion (fewer male births), whereas four areas are classified into the group of increased birth proportion (more male births). In columns VII to IX of Table 1.6 are the sorted Z-values with associated province number and C.A.MAN classifier. One of the fine points of this analysis consists in the separation of extreme observations under a single "normal" binomial distribution from those coming from separate subpopulations. In this analysis evidence is found for the existence of subpopulations with a live-birth male proportion different from $\lambda = 0.513$. A geographical characterization is given in Figure 1.17. Most provinces show a sex ratio at birth of medium level. There are 4 provinces with high sex ratio, namely Phrae, Samutprakhan, Rayong, and Ratchaburi. There are five provinces with a low sex ratio, namely Yala, Chiangrai, Chumphon, Lampang and Mahasarakham. Though there exist speculations about reasons for these deviations, none seems convincing enough to formulate any hypothesis.* Instead, these deviating provinces should be monitored carefully in future times.

1.7 Missing covariates and mixture models

We consider the situation of a simple regression model in which a response y is related to a covariate x. It is assumed further that they are related through a simple straight line model: $y \mid x$ is normally distributed $N(\alpha + \beta x, \sigma^2)$. The associated picture — usually also available in any introductory statistics book — is given in Figure 1.18. Three values of x, namely $x = 0$, $x = 3$, and $x = 6$, are used. Suppose further that 900 observations of pairs (y_i, x_i) are available with 300 placed at each of the three x-values. Then, the sample version of the population graph will be very similar to the one given in Figure 1.18. Now, suppose that the covariate x has *not* been observed, either because it is an *unknown covariate* (a covariate whose relationship to the response has not yet been detected) or it is too *complex* to measure (too invasive for the patient, too costly, etc.). Then, what are the consequences in terms of the distributional model? This is worked out in Figure 1.19. Since the data cannot be broken up with respect to x, one is left with the marginal distribution of the response y, which leads in a natural way to a 3-component mixture of normal distributions.

---

*Personal communication with the members of the Faculty of Public Health at Mahidol University, Bangkok, Thailand.

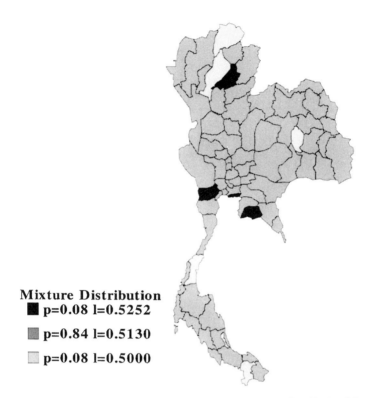

Figure 1.17. Estimated heterogeneity in live-birth ratio data for Thailand based on Census Data of 1990 (map was constructed using the classification rule (1.14) with P estimated by means of C.A.MAN).

A similar situation has been discussed in Example 1.6 where the most important covariate (effect covariate water-dip vs. skin-so-soft dip as flea repellent) had been ignored and a 2-component mixture of Poisson distributions could be observed. The estimated mixture component led then to a reconstruction of the missing covariate. However, it is not surprising that this reconstruction can only be complete up to a certain degree, and a certain loss of information necessarily occurs. Nevertheless, the occurrence of a mixture distribution can be viewed as *indicative* for a left-out covariate. This idea appears in various sciences under different names, such as *residual confounding* in epidemiology (for an application see Luoto *et al.* 1994), *frailty* in clinical trials, or *proneness* in traffic accident research. We will refer to this phenomenon as *latent* or *unobserved heterogeneity* (see also Heckman and Singer 1985).

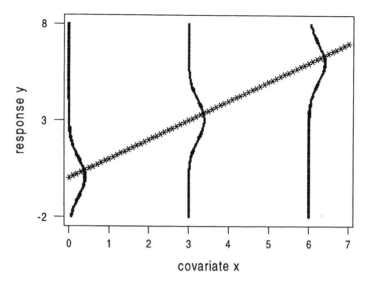

Figure 1.18. Hypothetical regression model $y = \alpha + \beta x + \varepsilon$ with $\alpha = 0$, $\beta = 1$, and normal error ε with mean 0 and variance 1.

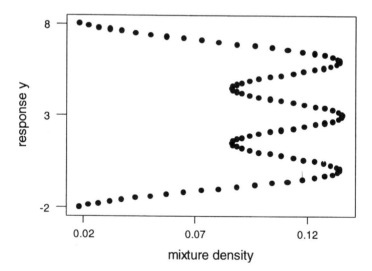

Figure 1.19. Marginal distribution of y with margin taken over the ignored covariate x: mixture of normals with 3 components.

CHAPTER 2

Theory of nonparametric mixture models

2.1   The likelihood and its properties

The strong results of nonparametric mixture distributions are based on the fact that the log-likelihood l is a *concave* functional on the set of *all* discrete probability distributions Ω. It is very important to distinguish between the set Ω of all discrete distributions and the set Ω_k of all distributions with a fixed number of k support points (subpopulations). The latter set is *not* convex.

Let P and Q be two discrete probability distributions in Ω. Then, with $\alpha \in [0, 1]$ the convex combination $(1 - \alpha)P + \alpha Q$ is also in Ω. This establishes that Ω is a convex set. The following example shows that Ω_k is *not* convex.

Example 2.1: Let

$$P = \begin{pmatrix} \lambda_1 \lambda_2 \\ p_1 p_2 \end{pmatrix} \text{ and } Q = \begin{pmatrix} \mu_1 \mu_2 \\ q_1 q_2 \end{pmatrix} \in \Omega_2$$

with *at least* $\lambda_1 \neq \mu_1$ or $\lambda_2 \neq \mu_2$. Then, for $\alpha \in (0, 1)$ the convex combination $(1 - \alpha)P + \alpha Q$ is *not* in Ω_2.

Lemma 2.1 The log-likelihood $l(P)$ is concave in Ω.

Proof. $l((1 - \alpha)P + \alpha Q) = \Sigma_i \log \mathrm{f}(x_i, (1 - \alpha)P + \alpha Q) =$

$$\Sigma_i \log \{(1 - \alpha)\, \mathrm{f}(x_i, P) + \alpha\, \mathrm{f}(x_i, Q)\} \geq (1 - \alpha)\, \Sigma_i \log \mathrm{f}(x_i, P)$$

$$+ \alpha\, \Sigma_i \log \mathrm{f}(x_i, Q) = (1 - \alpha)\, l(P) + \alpha\, l(Q),$$

where we have only used that the natural logarithm $\log(x)$ is a concave function of x. This completes a simple proof.

Definition 2.1 Any $\hat{P}$ which is maximizing $l\,(P)$ is called a *nonparametric maximum likelihood estimate* (NPMLE).[*]

2.2 The directional derivative and the gradient function

The major tool for achieving characterizations and algorithms is the *directional derivative at P into the direction Q*, for both P and Q in Ω:

$$\Phi(P, Q) = \lim_{\alpha \to 0} \frac{l((1-\alpha)P + \alpha Q) - l(P)}{\alpha} = \sum_{i=1}^{n} \frac{f(x_i;Q) - f(x_i;P)}{f(x_i;P)}.$$

The following inequality follows directly from the concavity of the log-likelihood.

Lemma 2.2 For any P and Q in Ω,

$$\Phi(P, Q) \geq l(Q) - l(P) \tag{2.1}$$

Proof. It follows from the concavity of l that

$$l((1-\alpha)\,P + \alpha Q) \geq l(P) + \alpha(l(Q) - l(P))$$

or,

$$\frac{l((1-\alpha)P + \alpha Q) - l(P)}{\alpha} \geq l(Q) - l(P).$$

Taking the limit $\alpha \to 0$ gives the desired result (2.1).

In particular, for one-point mass Q_λ at λ (the vertex of the simplex), the directional derivative is:

$$D_P(\lambda) = \Phi(P, Q_\lambda) = \sum_{i=1}^{n} \frac{f(x_i;\lambda) - f(x_i;P)}{f(x_i;P)} = \sum_{i=1}^{n} \frac{f(x_i;\lambda)}{f(x_i;P)} - n.$$

[*]The name goes back to Laird (1978).

Lemma 2.3

$$\sup_\lambda D_P(\lambda) \geq l(\hat{P}) - l(P) \qquad (2.2)$$

Proof. It is sufficient to show that $\sup_\lambda D_P(\lambda) \geq \sup_Q \Phi(P, Q)$. It is $Q = \Sigma_j p_j Q_{\lambda_j}$ for some $p_j \geq 0$ and $p_1 + \ldots + p_k = 1$, and $\lambda_1, \ldots, \lambda_k$. Thus, $\Phi(P, Q) = \Sigma_j p_j \Phi(P, Q_{\lambda_j})$. Because $D_P(\lambda) = \Phi(P, Q_\lambda)$ we achieve that $\Phi(P, Q) = \Sigma_j p_j \Phi(P, Q_{\lambda_j}) \leq \Sigma_j p_j \sup_\lambda \Phi(P, Q_\lambda) = \sup_\lambda D_P(\lambda)$. The right-hand side is independent of Q, therefore, taking the supremum on the left-hand side gives the desired result: $\sup_\lambda D_P(\lambda) \geq \sup_Q \Phi(P, Q)$.

The determining part in this directional derivative, namely,

$$\frac{1}{n} \sum_{i=1}^{n} \frac{f(x_i;\lambda)}{f(x_i;P)}$$

is called the *gradient function* and denoted by $d(\lambda, P)$. For the gradient function, the result (2.2) takes the form

$$\sup_\lambda d(\lambda, P) - 1 \geq \frac{1}{n} l(\hat{P}) - \frac{1}{n} l(P) \qquad (2.3)$$

The inequality (2.3) is very helpful in determining how far a candidate P_{cand} is from the maximum likelihood estimate and might be useful in constructing a suitable stopping rule for an algorithm.

2.3 The general mixture maximum likelihood theorem

With these preparations it is now quite easy to achieve the mixture maximum likelihood theorem. We have the general *mixture maximum likelihood theorem* (Lindsay 1983a, b; Böhning 1982):

Theorem 2.1:

(a) $\hat{P}$ is NPMLE if and only if $D_{\hat{P}}(\lambda) \leq 0$ for all λ or, if and only if

$$d(\lambda, \hat{P}) = \frac{1}{n} \sum_{i=1}^{n} \frac{f(x_i;\lambda)}{f(x_i;\hat{P})} \leq 1$$

for all λ in the parameter space.

(b) $D_{\hat{P}}(\lambda) = 0$ (or equivalently d$(\lambda, \hat{P}) = 1$) for all support points λ of $\hat{P} = \begin{pmatrix} \hat{\lambda}_1 \dots \hat{\lambda}_k \\ \hat{p}_1 \dots \hat{p}_k \end{pmatrix}$.

Proof.

Part (a). If $\hat{P}$ is NPMLE, then

$$\frac{l((1-\alpha)\hat{P} + \alpha Q) - l(\hat{P})}{\alpha} \le 0$$

for all Q in Ω, $\sup_Q \Phi(\hat{P}, Q) \le 0$. Since $\sup_Q \Phi(\hat{P}, Q) = \sup_\lambda D_{\hat{P}}(\lambda)$, we have the desired result. If $\sup_\lambda D_{\hat{P}}(\lambda) \le 0$, it follows from (2.2) that $\hat{P}$ is NPMLE.

Part (b). $0 = \Phi(\hat{P}, \hat{P}) = \Sigma_j \hat{p}_j \Phi(\hat{P}, Q_{\hat{\lambda}_j})$ and $\Phi(\hat{P}, Q_{\hat{\lambda}_j}) \le 0$ for all $\hat{\lambda}_j$. This induces $\Phi(\hat{P}, Q_{\hat{\lambda}_j}) = 0$, for all $\hat{\lambda}_j$ with $\hat{p}_j > 0$ and completes the proof.

Remark. The general mixture maximum likelihood theorem had some predecessors. Simar (1974) considered solely the case of Pois-son mixtures and Jewell (1982) the situation of mixtures of expo-nential distributions.

Corollary 2.1 Let $l(p) = l(p_1, p_2, \dots, p_m)$ be any concave and differentiable function on the finite-dimensional probability simplex $\Delta = \{p = p_1 e_1 + \dots + p_m e_m \mid p_j \ge 0$ for $j = 1, \dots, m$ and $p_1 + \dots + p_m = 1\}$ with e_j being the vector having only 0s and exactly one 1 at the jth position (*vertex*). Let $\hat{p} > 0$. Then, $\hat{p}$ is maximizing l in Δ if and only if,

$$\frac{\partial l}{\partial p_j}(\hat{p}) = \nabla l(\hat{p})^T \hat{p} \text{ for all } j = 1, \dots, m.$$

Proof. The proof follows along the lines of the general mixture maximum likelihood theorem and from the fact that the directional derivative $\Phi(p, q)$ can be written as $\nabla l(p)^T q - \nabla l(p)^T p$ for any $p, q \in \Delta$, and in particular, $\Phi(p, e_j) = \nabla l(p)^T e_j - \nabla l(p)^T p$ for any vertex direction.

Corollary 2.2 (Lindsay 1981)

Suppose that f(x_i, λ) — as a function of λ — has a *unique* mode for all x_i which lies in the interval $[x_{min}, x_{max}]$, where x_{min} and x_{max} are the minimum and maximum of the observed data $x_1, \dots, x_n$, respectively.

Then, $\hat{P}$ can only have support points in the interval $[x_{min}, x_{max}]$.

Proof. We write

$$d(\lambda, \hat{P}) = \frac{1}{n}\sum_{i=1}^{n}\frac{f(x_i;\lambda)}{f(x_i;\hat{P})} = \frac{1}{n}\sum_{i=1}^{n}a_i f(x_i;\lambda)$$

with $a_i = 1/f(x_i, \hat{P})$. According to Theorem 2.1 support points of $\hat{P}$ are also modes of $d(\lambda, \hat{P})$. Now since all n functions $f(x_i, \lambda)$ are strictly increasing for $\lambda \leq x_{min}$ and strictly decreasing for $\lambda \geq x_{max}$, the linear sum

$$\frac{1}{n}\sum_{i=1}^{n}a_i f(x_i;\lambda)$$

also has this property, e.g., $d/d\lambda\, d(\lambda, \hat{P}) > 0$ for $\lambda < x_{min}$ and $d/d\lambda\, d(\lambda, \hat{P}) < 0$ for $\lambda > x_{max}$. Thus, $\hat{P}$ can have *no* points of support $((d/d\lambda)d(\lambda, \hat{P}) = 0)$ outside the interval $[x_{min}, x_{max}]$. This ends the proof.

Example 2.2: For many densities the assumption of Corollary 2.2 is trivially fulfilled. Consider the normal density

$$f(x, \lambda) = (2\pi\sigma^2)^{-1/2}\exp\left(-\frac{1}{2}(x-\lambda)^2/\sigma^2\right)$$

which is maximized for $\lambda = x$. The Poisson $f(x, \lambda) = \exp(-\lambda)\,\lambda^x/x!$ is also maximized by $\lambda = x$ (consider $\log f(x, \lambda) = -\lambda + x \log \lambda - \log(x!)$). However, consider the binomial

$$f(x, \lambda, N) = \binom{N}{x}\lambda^x(1-\lambda)^{N-x} .$$

The binomial density is maximized for $\lambda = x/N$. Consequently, the NPMLE will have points of support only in the interval $[\min\{x_i/N_i\}, \max\{x_i/N_i\}]$. Similarly, if the Poisson kernel $f(x, \lambda, e) = \exp(-\lambda e)\,(\lambda e)^x/x!$ is considered, its mode is given by x/e and the NPMLE can have only support points in the interval $[\min\{x_i/e_i\}, \max\{x_i/e_i\}]$.

Corollary 2.2 has implications in that there is no need to search for the NPMLE outside the range of observed data. This reduces the computational burden enormously.

2.4   Applications of the theorem

One of the potential applications of the theorem is to check the optimality of a candidate P. In many cases, it will be of interest if the
situation of parameter *homogeneity* or *heterogeneity* is true. Again,
often the maximum likelihood estimate under *homogeneity* is known,
often simply the arithmetic mean $\bar{x}$.

Example 2.3: Let $f(x, \lambda) = \exp(-\lambda)\,\lambda^x/x!$, the Poisson density. Then, $d(\lambda, \bar{x}) = (1/n)\Sigma_i\,f(x_i, \lambda)/f(x_i, \bar{x}) = \exp(\bar{x} - \lambda)\,(1/n)\Sigma_i\,(\lambda/\bar{x})^{x_i}$. To demonstrate the
usefulness of the approach, we look at a simulated data set of size $n = 100$ from a homogeneous Poisson distribution with $\lambda = 5$. The counts are
given in Table 2.1.

x	1	2	3	4	5	6	7	8	9	10
frequency	2	10	17	20	19	12	10	4	4	2

Table 2.1. Simulated data set of Poisson counts of size 100, $\lambda = 5$

It becomes quite clear from Figure 2.1 that there is homogeneity. The
gradient function is bounded by 1 and the bound is sharp for $\lambda = \bar{x} = 4.78$. Thus, $\bar{x}$ is the NPMLE.

Example 2.4: Frequently a specific form of Poisson kernel is of interest.
Observed death counts x are related to expected death counts e, where
e is computed as $n_1\mu_1 + n_2\mu_2 + \ldots + n_J\mu_J$, where n_j is the size of age-
stratum j in the study population and μ_j is the mortality rate in age-
stratum j in a *reference population* (for details see Ahlboom 1995). Consequently, a specific form of the Poisson kernel is required:

$$f(x, \lambda, e) = \exp(-\lambda e)\,(\lambda e)^x/x! \qquad (2.4)$$

where λ is the expected value of the standardized mortality ratio SMR,
defined as SMR $= x/e$. It is easy to see that the maximum likelihood
estimator of λ under homogeneity is given as $\hat{\lambda} = \Sigma_i\,x_i/\Sigma_i\,e_i$.

Then,

$$d(\lambda, \hat{\lambda}) = \frac{1}{n}\Sigma_i f(x_i, \lambda, e_i)/f(x_i, \hat{\lambda}, e_i) = \frac{1}{n}\Sigma_i(\lambda/\hat{\lambda})^{x_i}\exp((\hat{\lambda} - \lambda)e_i).$$

We look at a realistic data set of SMR-values published by Martuzzi and
Hills (1995). The authors studied perinatal mortality in a health region
in England consisting of 515 areas.

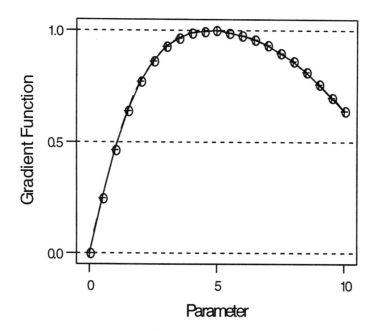

Figure 2.1. Gradient function d(λ, $\bar{x}$) for simulated data set from a homogeneous Poisson distribution with mean 5.

Given that we are willing to accept that a Poisson distribution of the form (2.4) is correct in each region, then Figure 2.2 clearly identifies a violation of *homogeneity*. In fact, the likelihood can be improved upon if mass is moved towards the parameters $\lambda = 2$ and $\lambda = 0.5$. We will come back to this issue at some later point.

Example 2.5: Consider the normal kernel with known variances $\sigma^2 = 1$, $f(x, \lambda) = (2\pi)^{-1/2} \exp(-(1/2)(x - \lambda)^2)$. Then,

$$d(\lambda, \bar{x}) = \frac{1}{n}\Sigma_i f(x_i, \lambda)/f(x_i, \bar{x}) = \frac{1}{n}\Sigma_i \exp\left(\frac{1}{2}(x_i - \bar{x})^2 - \frac{1}{2}(x_i - \lambda)^2\right)$$

$$= \frac{1}{n}\Sigma_i \exp\left\{x_i(\lambda - \bar{x}) - \frac{1}{2}(\lambda^2 - \bar{x}^2)\right\} = \exp\left\{\frac{1}{2}(\bar{x}^2 - \lambda^2)\right\}\frac{1}{n}\Sigma_i \exp\{x_i(\lambda - \bar{x})\}.$$

Remark. One should note that these gradient function plots have been achieved without any large computational burden. In fact, they can be constructed with any simple package able to execute macros,

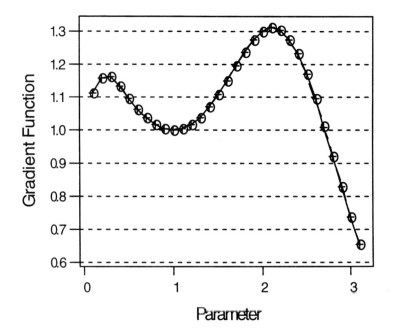

Figure 2.2. Gradient function for SMR-Poisson kernel in the data set of Martuzzi and Hills of 515 SMR-data of the North West Thames health region in England.

such as the package MINITAB which has been used to compute the gradient functions in Examples 2.2 and 2.3.

2.5 Existence and number of support points of the NPMLE

Suppose that the sample incorporates replications. In other words, there are only y_1, y_2, ..., y_m different values, each occurring w_1, w_2, ..., w_m times. The log-likelihood takes the form

$$l(P) = \Sigma_i \ w_i \ \log \ f(y_i, P).$$

The results in this section will strongly depend on the geometry of convex sets. Consider the m-dimensional set

$$\Gamma = \{(f(y_1, \lambda), \ ...., \ f(y_m, \lambda))^T \ | \ \lambda \in \Lambda\}.$$

Let also conv(Γ) denote the *convex hull of* Γ, i.e.,

$$\text{conv}(\Gamma) = \{\ \Sigma_j\ p_j\,\mathbf{f}_j\ \mid\ \mathbf{f}_j \in \Gamma,\ p_i \geq 0,\ p_1 + p_2 + \ldots + p_J = 1\ \}.$$

If Γ is a closed and bounded set, then conv(Γ) is also closed and bounded (Eggleston 1966, Theorem 10). This guarantees that the continuous function $\phi(\mathbf{f}) = \Sigma_i\ w_i\ \log f_i$ attains its maximum on conv(Γ). Note that ϕ is strictly concave, thus there is a unique $\hat{\mathbf{f}}$ to maximize ϕ. However, there might be several $\hat{P}$ with $(f(y_1, \hat{P}), \ldots, f(y_m, \hat{P}))^T = \hat{\mathbf{f}}$.

Now, a well-known theorem of Carathéodory (Eggleston 1966, Theorem 18) establishes that any point $\mathbf{f}$ out of the convex hull conv(Γ) can be represented as a finite convex combination $p_1\mathbf{f}_1 + \ldots + p_J\mathbf{f}_J$, with $\mathbf{f}_j \in \Gamma$, $p_i \geq 0$, $p_1 + p_2 + \ldots + p_J = 1$ *and* $J \leq m + 1$. If $\mathbf{f}$ is on the boundary, then $J \leq m$. We summarize these results in the following Theorem 2.2.

Theorem 2.2 (Lindsay 1982):

(a) If Γ is closed and bounded, then a NPMLE $\hat{P}$ exists.

(b) $\hat{P}$ has at most m points of support.

Theorem 2.2 has important implications. It guarantees the *discreteness* of the NPMLE in full generality. Furthermore, it gives a bound on the maximum number of necessary support points. This bound is the sample size n, if all data points are different from each other. However, if there are many replications this bound is largely reduced. To demonstrate, let us consider Example 2.3 with $n = 100$, but $m = 10$. Here, no more than 10 support points for the NPMLE are necessary (we know, of course, that it has only *one* support point, in this case). In practice, the bound for the number of support points given in Theorem 2.2 is seldom sharp, e.g., there will often be fewer support points required than the bound indicates.

2.6   A case study on Simar's accident data

In this section we illustrate the difficulties involved in computing the nonparametric maximum likelihood estimate. In the illustration we use the gradient function as a diagnostic device to detect if a given estimate of P is the NPMLE or not. Simar (1976) provided one of the pioneering papers on the NPMLE for mixtures of Poisson distributions. In the paper, count data were studied which were accident counts of

y_i	0	1	2	3	4	5	6	7
frequency w_i	7840	1317	239	42	14	4	4	1

Table 2.2. Accident data of Thyrion (1960) used by Simar (1974)

accident insurance policies reporting exactly y_i claims during a particular year for $n = 9461$ policies issued by La Royal Belge Insurance
Company. The data (see Table 2.2) go back to Thyrion (1960) and have
been used on various occasions in the literature, including Carlin and
Louis (1996, p. 72).
    The estimate of P given by Simar is

$$\hat{P} = \begin{pmatrix} 0.089 & 0.580 & 3.176 & 3.669 \\ 0.7600 & 0.2362 & 0.0037 & 0.0002 \end{pmatrix}$$

with associated log-likelihood of -5341.5310. This estimator has been
reported to be the NPMLE on various occasions including Carlin and
Louis (1996, p. 74, Table 3.2). As it turns out, this is *not* the NPMLE.
An inspection of the gradient function as given in Figure 2.3 shows
that it attains values above 1 in the neighborhood of 0. Using one of
the reliable converging algorithms of the next chapter it can be worked
out that the NPMLE is in this case

$$\hat{P} = \begin{pmatrix} 0. & 0.3356 & 2.5454 \\ 0.4184 & 0.5730 & 0.0087 \end{pmatrix}$$

with associated log-likelihood of -5340.7040. Note that not only this
likelihood is larger, also the gradient function is well below 1, as Figure
2.4 demonstrates. The general problem in finding maximum likelihood
estimates for mixture models lies in the *flatness* of the associated
likelihood surfaces. For a demonstration we look at the conventional
log-likelihood under homogeneity

$$l_0(\lambda) = \Sigma_i \, w_i \, \log \, f(y_i, \lambda) \tag{2.5}$$

where the sum is from 1 to m. This likelihood is maximized for $\bar{x} = 0.2144$. We take this as the likelihood for comparison. Now, we consider

$$l_1(\lambda) = \Sigma_i w_i \log\{0.4184 \, f(y_i, 0) + 0.5730 \, f(y_i, \lambda) \tag{2.6}$$
$$+ \, 0.0087 \, f(y_i, 2.5454)\}$$

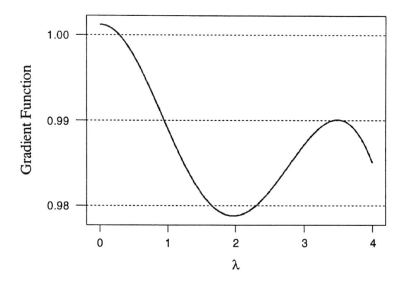

Figure 2.3. Gradient function for the accident data of Simar (1976) and the estimator of P given by Simar.

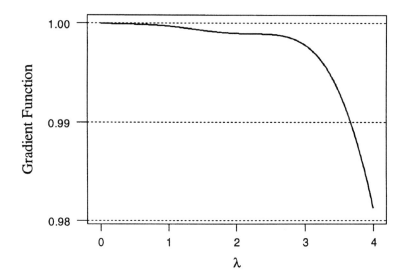

Figure 2.4. Gradient function for the accident data of Simar (1976) and the nonparametric maximum likelihood estimator of P.

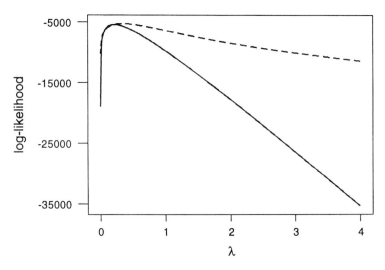

Figure 2.5. Graphs of the log-likelihoods $l_0(\lambda)$ and $l_1(\lambda)$ as defined in (2.5) and (2.6), respectively.

that is, wo conoidor tho mixturc log likclihood as a function of the second component mean. The corresponding graph is provided in Figure 2.5. Evidently, both log-likelihoods show considerable curvatures. This changes if instead of $l_1(\lambda)$ the log-likelihood $l_2(\lambda)$ defined as

$$l_2(\lambda) = \Sigma_i w_i \log\{0.4184\ f(y_i, 0) + 0.5730\ f(y_i, 0.3356) \atop + 0.0087\ f(y_i, \mu)\}, \qquad (2.7)$$

that is, we consider the mixture log-likelihood as a function of the third component. The corresponding graph is provided in Figure 2.6. It appears as if there is almost no curvature in the log-likelihood $l_2(\lambda)$. This becomes even more clear, if we look at mixture log-likelihood as a function of $\lambda_1 = \lambda$ and $\lambda_2 = \mu$:

$$l_3(\lambda) = \Sigma_i w_i \log\{0.4184\ f(y_i, 0) + 0.5730\ f(y_i, \lambda) \atop + 0.0087\ f(y_i, \mu)\}. \qquad (2.8)$$

There appears to be a *flat ridge* in the direction of μ, which is creating the problem here.

This case study indicates what kind of problems might occur when dealing with mixture likelihoods. Algorithms are required which can reliably maximize flat likelihood surfaces. We will deal with these in the next chapter.

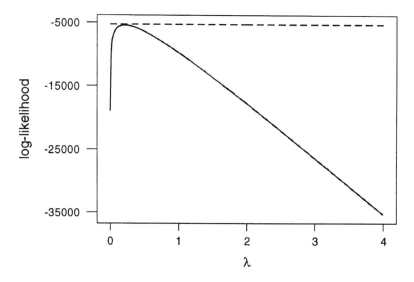

Figure 2.6. Graphs of the log-likelihoods $l_0(\lambda)$ and $l_2(\lambda)$ as defined in (2.5) and (2.7), respectively.

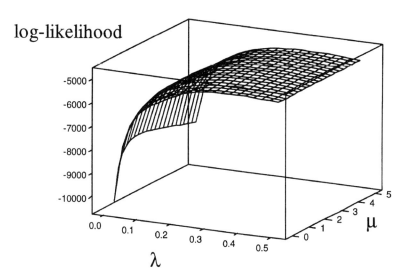

Figure 2.7. Log-likelihood $l_3(\lambda, \mu)$ as a function of $\lambda = \lambda_2$ and $\mu = \lambda_3$ for fixed values of $\lambda_1 = 0$ and the optimal weights p_1, p_2, and p_3.

CHAPTER 3

Algorithms

3.1 The vertex direction method (VDM)

We have seen in Chapter 2 not only how important the concept of a directional derivative is in deriving characterizations, but that it is also fundamental in developing suitable algorithms for constructing the NPMLE. Historically, the vertex direction method (VDM) is of interest. In this method convex combinations $(1 - \alpha)P + \alpha Q_\lambda$ are considered, for which the log-likelihood increase $l((1 - \alpha)P + \alpha Q_\lambda) - l(P)$ (as a function of the step-length α and the vertex direction Q_λ) is desired to be made as large as possible. A first-order approximation $\alpha \, \partial\phi/\partial\alpha \, (0) = \alpha \, D_P(\lambda)$ (with $\phi(\alpha) = l((1 - \alpha)P + \alpha Q_\lambda) - l(P)$) of this difference leads to the maximization of $D_P(\lambda)$ in λ. Having found a vertex direction with maximum increase one can choose an appropriate (optimal) step-length α to achieve an update $(1 - \alpha)P + \alpha Q_\lambda$.

Example 3.1: Let us consider a simple situation. Let $\Lambda = \{\lambda_1, \lambda_2, \lambda_3\}$, then Ω is the 2-dimensional simplex $\{p = (p_1, p_2, p_3)^T \mid p_i \geq 0, p_1 + p_2 + p_3\}$. Note that $e_1 = (1, 0, 0)^T$, $e_2 = (0, 1, 0)^T$, $e_3 = (0, 0, 1)^T$, the three vertices of the simplex. Figure 3.1 shows the directions of movement for the VDM in this case.

Algorithm 3.1: Let P_0 be any initial value, for example $P_0 = Q_{\bar{x}}$.

Step 1. Find λ_{max} such that $D_P(\lambda_{max}) = \sup_\lambda D_P(\lambda)$.

Step 2. Find α_{max} such that $l((1 - \alpha_{max}) P_n + \alpha_{max} Q_{\lambda_{max}}) = \sup_\alpha l((1 - \alpha)P_n + \alpha \, Q_{\lambda_{max}})$.

Step 3. Set $P_{n+1} = (1 - \alpha_{max}) P_n + \alpha_{max} Q_{\lambda_{max}}$, $n = n + 1$ and go to step 1.

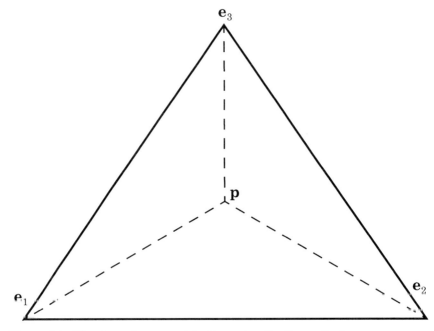

Figure 3.1. Directions of movement for the vertex direction method.

Theorem 3.1: Let (P_n) be any sequence created by Algorithm 3.1.

Then

$$l(P_n) \to l(\hat{P}) \text{ monotonously.}$$

Proof. By construction, the sequence $l(P_n)$ is monotone non-decreasing, e.g., $l(P_{n+1}) \geq l(P_n)$ for all n. Since the sequence is bounded above, it has to converge to some value L, say. Now, suppose $l(\hat{P}) - L = \varepsilon > 0$. Consider the second-order Taylor representation

$$l((1 - \alpha)P_n + \alpha Q_{\lambda_{max}}) - l(P_n)$$

$$= \alpha D_{P_n}(\lambda_{max}) + \frac{1}{2}\alpha^2 \frac{\partial^2}{(\partial \alpha)^2} l((1 - \alpha)P_n + \alpha Q_{\lambda_{max}})\big|_{\alpha = \alpha^*}$$

where $\alpha^* \in [0, 1]$. By means of (2.2) we have that

$$D_{P_n}(\lambda_{max}) \geq l(\hat{P}) - l(P_n) \geq l(\hat{P}) - L = \varepsilon > 0.$$

Since $D_{P_n}(\lambda_{max})$ is bounded away from 0, α_{max} will be also bounded away from 0, by some bound α_0, say. Then, let K be the bound for the second derivative $\partial^2/(\partial\alpha)^2\, l((1-\alpha)\,P_n + \alpha Q_{\lambda_{max}})$ and let α_δ be some $\alpha > 0$ for which $\alpha K \geq -\delta$ for given $\delta = \varepsilon/2$. Then, for all $\alpha \in [0, \min\{\alpha_0, \alpha_\delta\}]$ we have

$$l((1-\alpha)P_n + \alpha Q_{\lambda_{max}}) - l(P_n) \geq \alpha\,(\varepsilon - \varepsilon/2) = \alpha\varepsilon/2.$$

since $l(P_{n+1}) - l(P_n) \geq l((1-\alpha)P_n + \alpha Q_{\lambda_{max}}) - l(P_n) \geq \alpha\varepsilon/2$. Now choose, for example, $\alpha = \min\{\alpha_0, \alpha_\delta\}$; then

$$l(P_{n+1}) - l(P_0) = \sum_{i=0}^{n} l(P_{i+1}) - l(P_i) \geq (n+1)\alpha\varepsilon/2\,,$$

which converges to $+\infty$ if n converges to $+\infty$, which is impossible since the log-likelihood l is bounded above. This ends the proof.

Ideas on improvement. Since $\alpha\, D_P(\lambda)$ is only a rough approximation 4747of $\phi(\alpha)$ (and also in its size dependent on the distance of the direction to the current P), it is more appropriate to look at second-order approximations of

$$\phi(\alpha) = l((1-\alpha)P + \alpha Q_\lambda) - l(P) \approx \alpha\, D_P(\lambda) + \tfrac{1}{2}\,\alpha^2\, D_P^{(2)}(\lambda),$$

the latter being maximized for $\alpha_{max} = -D_P(\lambda)/D_P^{(2)}(\lambda)$. If $\alpha = \alpha_{max}$ is replaced in $\alpha\, D_P(\lambda) + \tfrac{1}{2}\alpha^2\, D_P^{(2)}(\lambda)$, then we achieve $\Delta_P(\lambda) = -\tfrac{1}{2}[D_P(\lambda)]^2/D_P^{(2)}(\lambda)$. Here,

$$D_P^{(2)}(\lambda) = \frac{\partial^2}{(\partial\alpha)^2}l((1-\alpha)P + \alpha Q_\lambda)\big|_{\alpha=0} = -\sum_{i=1}^{n} \frac{[f(x_i, \lambda) - f(x_i, P)]^2}{f(x_i, P)^2}.$$

Note that if $D_P(\lambda)$ is written as $\Sigma_i A_i$, $A_i = [f(x_i, \lambda) - f(x_i, P)]/f(x_i, P)$, then $D_P^{(2)}(\lambda) = -\Sigma_i A_i^2$ and $\Delta_P(\lambda) = \tfrac{1}{2}[\Sigma_i A_i]^2/\Sigma_i A_i^2$. An alternative VDM would choose the vertex λ_{max} according to the maximal value of $\Delta_P(\lambda)$. A convergence proof is along the lines of the proof for Theorem 3.1.

3.2   The vertex exchange method (VEM)

The VDM is usually slow in its convergence behavior. A faster method (also now used as the standard method in the package C.A.MAN) is the vertex exchange method (VEM). The basic idea here is to exchange *good* vertex directions for bad ones which are already in support[*] of the current mixing distribution. The VEM is defined as $P + \alpha P(\lambda^*)\{Q_\lambda - Q_{\lambda^*}\}$, where $P(\lambda^*)$ is the weight of the *bad* support point λ^* and α in [0, 1] is a step-length. Good and bad support points are identified again by means of the directional derivative. Again, one tries to optimize the gain in the log-likelihood $l(P + \alpha\, P(\lambda^*)\,\{Q_\lambda - Q_{\lambda^*}\}) - l(P)$. To understand the mechanism of the VEM in detail, let us consider the following example.

Example 3.2: We are in the situation of Example 3.1. Suppose, the procedure is at $p = (1/3, 1/3, 1/3)^T$ and wants to move mass from e_1 to e_2. The VEM-step is here $p + \alpha p_1\{e_2 - e_1\}$. For $\alpha = 0$, we yield p and for $\alpha = 1$ we are at $p + p_1\{e_2 - e_1\} = (0, 2/3, 1/3)^T$. This point lies in the *edge* of the simplex, namely, the one which connects e_2 with e_1. See Figure 3.2. Obviously, the VEM moves parallel to the edges of the simplex.

Now, consider a first-order approximation of this difference: $\alpha P(\lambda^*)\{D_P(\lambda) - D_P(\lambda^*)\}$. Clearly, this is maximized if $D_P(\lambda)$ is maximized in λ and $D_P(\lambda^*)$ is minimized in the support of P. Choosing an appropriate step-length completes the VEM. For details or different methods, see Böhning (1989, 1995) or Lesperance and Kalbfleisch (1992).

Algorithm 3.2: Let P_0 be any initial value, for example, $P_0 = Q_{\bar{x}}$.

Step 1. Find λ_{max} such that $D_{P_n}(\lambda_{max}) = \sup_\lambda D_{P_n}(\lambda)$ and λ_{min} such that $D_{P_n}(\lambda_{min}) = \min\{D_{P_n}(\lambda) \mid \lambda \in \text{support of } P_n\}$.

Step 2. Find α_{max} such that

$$l(P_n + \alpha_{max}P_n(\lambda_{min})\{Q_{\lambda_{max}} - Q_{\lambda_{min}}\})$$
$$= \sup_\alpha l(P_n + \alpha P_n(\lambda_{min})\{Q_{\lambda_{max}} - Q_{\lambda_{min}}\}).$$

Step 3. Set $P_{n+1} = P_n + \alpha_{max} P_n(\lambda_{min})\{Q_{\lambda_{max}} - Q_{\lambda_{min}}\}$, $n = n + 1$ and go to step 1.

Lemma 3.1: Let λ_{min} be such that $D_P(\lambda_{min}) = \min\{D_P(\lambda) \mid \lambda \in \text{support of } P\}$. Then, $D_P(\lambda_{min}) \leq 0$.

Proof. Suppose that $D_P(\lambda_j) > 0$ for all support points λ_j of P. This also implies that $\Sigma_j D_P(\lambda_j) p_j > 0$, where p_j are the positive weights of the associated support points. Then, because of the linearity of the directional derivative,

[*]The *support* of a probability measure is the set of points receiving *positive* weight.

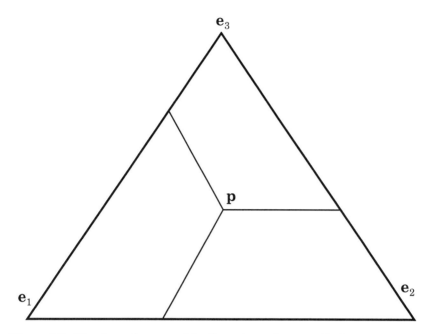

Figure 3.2. Directions of movement for the vertex exchange method.

$$\Sigma_j D_P(\lambda_j)p_j = \Sigma_j \Phi(P, Q_{\lambda_j})p_j = \Phi(P, \Sigma_j p_j Q_{\lambda_j}) = \Phi(P, P) = 0$$

which contradicts $\Sigma_j D_P(\lambda_j)\, p_j > 0$.

Theorem 3.2: Let (P_n) be any sequence created by Algorithm 3.2. Then

$$l(P_n) \rightarrow l(\hat{P}) \text{ monotonously.}$$

Proof. By construction, the sequence $l(P_n)$ is monotone non-decreasing, e.g., $l(P_{n+1}) \geq l(P_n)$ for all n. Since the sequence is bounded above, it has to converge to some value L, say. Now, suppose $l(\hat{P}) - L = \varepsilon > 0$. Consider the second-order Taylor representation

$$l(P_n + \alpha P_n(\lambda_{min})\{Q_{\lambda_{max}} - Q_{\lambda_{min}}\}) - l(P_n)$$

$$= \alpha P_n(\lambda_{min})\{D_{P_n}(\lambda_{max}) - D_{P_n}(\lambda_{min})\}$$

$$+ \frac{1}{2}\alpha^2 \frac{\partial^2}{(\partial\alpha)^2}l(P_n + \alpha P_n(\lambda_{min})\{Q_{\lambda_{max}} - Q_{\lambda_{min}}\})\big|_{\alpha = \alpha^*}$$

where $\alpha^* \in [0, 1]$. By means of (2.2) and Lemma 3.1 we have that

$$D_{P_n}(\lambda_{\max}) - D_{P_n}(\lambda_{\min}) \geq D_{P_n}(\lambda_{\max}) \geq l(\hat{P}) - l(P_n) \geq l(\hat{P}) - L = \varepsilon > 0.$$

Since $D_{P_n}(\lambda_{\max}) - D_{P_n}(\lambda_{\min})$ is bounded away from 0, $\alpha_{\max}$ will be also bounded away from 0, by some bound α_0, say. Then, let K be a bound for the second derivative $\partial^2/(\partial\alpha)^2 \, l(P_n + \alpha\, P_n(\lambda_{\min})\{Q_{\lambda_{\max}} - Q_{\lambda_{\min}}\})$ and let α_δ be some $\alpha > 0$ for which $\alpha K \geq -\delta$ for given $\delta = \varepsilon/2$. Then, for all $\alpha \in [0, \min\{\alpha_0, \alpha_\delta\}]$ we have

$$l(P_n + \alpha\, P_n(\lambda_{\min})\{Q_{\lambda_{\max}} - Q_{\lambda_{\min}}\}) - l(P_n)$$
$$\geq \alpha\, P_n(\lambda_{\min})\,(\varepsilon - \varepsilon/2) = \alpha\, P_n(\lambda_{\min})\varepsilon/2.$$

Consequently,

$$l(P_{n+1}) - l(P_n) = l(P_n + \alpha_{\max}\, P_n(\lambda_{\min})\{Q_{\lambda_{\max}} - Q_{\lambda_{\min}}\}) - l(P_n)$$
$$\geq l(P_n + \alpha\, P_n(\lambda_{\min})\{Q_{\lambda_{\max}} - Q_{\lambda_{\min}}\}) - l(P_n) \geq \alpha P_n(\lambda_{\min})\varepsilon/2.$$

Now choose, for example, $\alpha = \min\{\alpha_0, \alpha_\delta\}$, then

$$l(P_{n+1}) - l(P_0) = \sum_{i=0}^{n} (l(P_{i+1}) - l(P_i)) \geq (n+1)\alpha P_n(\lambda_{\min})\varepsilon/2,$$

which converges to $+\infty$ if n converges to $+\infty$, which is impossible since the log-likelihood l is bounded above. This ends the proof.

The VEM is a stable procedure and converges better than the VDM. Lesperance and Kalbfleisch (1992) consider an example in which the VDM needs 2177 iterations, whereas the VEM needs only 143.

The above algorithms need to be stopped at some iteration n with associated P_n close to $\hat{P}$. Recall that

$$D_{P_n}(\lambda_{\max}) \geq l(\hat{P}) - l(P_n).$$

Therefore, $D_{P_n}(\lambda_{\max}) < \varepsilon$ will guarantee that $l(\hat{P}) - l(P_n) < \varepsilon$. We will utilize this idea to create a *stopping rule*.

Stopping rule 3.1: Fix $\varepsilon > 0$ to some small number such as 0.01, 0.001, 0.0001, or 0.00001. Stop algorithm 3.1 or 3.2 if $D_{P_n}(\lambda_{\max}) < \varepsilon$.

Example 3.3: We are in the situation of Example 1.1 and we want to compare the convergence behavior of the VDM with the one of the VEM. In both cases optimal step-length is used and ε is set to 0.0001. The VDM needs 9901 iterations to meet the stopping rule and is doing this in 9 seconds. The VEM needs 475 iterations and is doing this in 1.5 seconds. This underlines the superiority of the VEM in terms of convergence behavior. More details on computational comparisons can be found in Böhning (1985, 1986).

Ideas on improvement. Since $\alpha P(\lambda^*)[D_P(\lambda) - D_P(\lambda^*)]$ is only a rough approximation of $\phi(\alpha)$, it is more appropriate to look at second-order approximations of

$$\phi(\alpha) = l(P + \alpha P(\lambda^*)[Q_\lambda - Q_{\lambda*}]) - l(P) \approx \alpha\phi'(\alpha)|_{\alpha = 0} \qquad (3.1)$$

$$+ \frac{1}{2}\alpha^2\phi''(\alpha)|_{\alpha = 0}$$

$$= \alpha P(\lambda^*)[D_P(\lambda) - D_P(\lambda^*)] + \frac{1}{2}\alpha^2\tilde{D}_P^{(2)}(\lambda, \lambda^*)$$

where

$$\tilde{D}_P^{(2)}(\lambda, \lambda^*) = \frac{\partial^2}{(\partial\alpha)^2}l(P + \alpha P(\lambda^*)[Q_\lambda - Q_{\lambda*}])|_{\alpha = 0}$$

$$= -P(\lambda^*)^2\Sigma_i[f(x_i, \lambda) - f(x_i, \lambda^*)]^2/f(x_i, P)^2.$$

The maximizing value for the quadratic in α is given by

$$\alpha_{\max} = -P(\lambda^*)[D_P(\lambda) - D_P(\lambda^*)]/\tilde{D}_P^{(2)}(\lambda, \lambda^*)$$

$$= \frac{\Sigma_i[f(x_i, \lambda) - f(x_i\lambda^*)]/f(x_i, P)}{\Sigma_i[f(x_i, \lambda) - f(x_i, \lambda^*)]^2/f(x_i, P)^2}/P(\lambda^*) \qquad (3.2)$$

Inserting $\alpha = \alpha_{\max}$ in (3.1) leads to

$$\frac{1}{2}\frac{\{\Sigma_i[f(x_i, \lambda) - f(x_i, \lambda^*)]/f(x_i, P)\}^2}{\Sigma_i[f(x_i, \lambda) - f(x_i, \lambda^*)]^2/f(x_i, P)^2} . \qquad (3.3)$$

To ensure that those vertices are chosen which are directions of ascent, we consider the positive root of (3.3) as search criterion

$$\Delta_P(\lambda, \lambda^*) := \frac{\Sigma_i[f(x_i, \lambda) - f(x_i, \lambda^*)]/f(x_i, P)}{\sqrt{\Sigma_i[f(x_i, \lambda) - f(x_i, \lambda^*)]^2/f(x_i, P)^2}}. \tag{3.4}$$

Note that for given λ^* and maximizing value of λ the value of α in (3.2) is non-negative: $\alpha_{max} \geq 0$.

Algorithm 3.3: Let P_0 be any initial value.

Step 0. Find λ_{min} such that $D_{P_n}(\lambda_{min}) = \min \{D_{P_n}(\lambda) \mid \lambda \in$ support of $P_n\}$.

Step 1. Find λ_{max} such that $\Delta_{P_n}(\lambda_{max}, \lambda_{min}) = \sup_\lambda \Delta_{P_n}(\lambda, \lambda_{min})$.

Step 2. Find λ_{min} such that $\Delta_{P_n}(\lambda_{max}, \lambda_{min}) = \min \{\Delta_{P_n}(\lambda_{max}, \lambda) \mid \lambda \in$ support of P_n\}.

Step 3. Set $P_{n+1} = P_n + \alpha_{max} P_n(\lambda_{min})\{Q_{\lambda max} - Q_{\lambda min}\}$, with α_{max} according to (3.2),* $n = n + 1$ and go to step 0.

Although Algorithm 3.3 is simple and converges quickly, a theoretical proof of convergence suffers from a lack of monotonicity in its construction, though in practice it is difficult to find counterexamples. This lack of monotonicity occurs because the quadratic approximation (3.1) might be below or above the log-likelihood difference. Nevertheless it is possible to develop a monotonic version of Algorithm 3.3. The key idea here lies in the fact that not only is $\phi(\alpha)$ *concave* in α, but so is $\phi''(\alpha)$. Now, consider the Taylor series development

$$\phi(\alpha) = \alpha\phi'(0) + \tfrac{1}{2} \alpha^2 \phi''(\overline{\alpha})$$

with some $\overline{\alpha} \in [0, 1]$. Therefore, it is possible to find a lower bound such that $\phi''(\overline{\alpha}) \geq \min\{\phi''(0), \phi''(1)\}$, since a concave function attains its minimum on an interval only at the interval ends. It follows that $\phi(\alpha) \geq \alpha\phi'(0) + \tfrac{1}{2} \alpha^2 \min\{\phi''(0), \phi''(1)\}$ for all α. The maximizing value of α is $\alpha_{max} = -\phi'(0)/\min\{\phi''(0), \phi''(1)\}$ and the quadratic takes the value $-\phi'(0)^2/\min\{\phi''(0), \phi''(1)\}$. It is not difficult to verify that $-\phi'(0)^2/\min\{\phi''(0), \phi''(1)\}$ is in our case

*If $\alpha_{max} > 1$, then it is truncated to 1.

$$\frac{\frac{1}{2}\{\Sigma_i[f(x_i, \lambda) - f(x_i, \lambda^*)]/f(x_i, P)\}}{M}$$

where $M = \min \{\Sigma_i [f(x_i, \lambda) - f(x_i, \lambda^*)]^2/f(x_i, P)^2, \Sigma_i [f(x_i, \lambda) - f(x_i, \lambda^*)]^2/f(x_i, P + P(\lambda^*)[Q_\lambda - Q_{\lambda^*}])^2\}$.

3.3   Step-length choices

In this section we consider *good* choices of the step-length α. Specific interest is devoted to monotonic procedures, since they guarantee a non-decreasing likelihood. Let, as in (3.1), $\phi(\alpha)$ denote

$$\phi(\alpha) = l(P + \alpha H) - l(P),$$

where H is the vertex direction or the vertex exchange direction.* Again, not only ϕ is concave but so is ϕ''. This will have important consequences.

We begin by offering another interpretation of the Newton–Raphson algorithm as an *area estimation* algorithm. Consider

$$A(\alpha) = \int_0^\alpha \phi''(t)\mathrm{d}t = \phi'(\alpha) - \phi'(0), \qquad (3.5)$$

the area above the curve ϕ'' from 0 to α. Specifically, for the line maximizer $\hat{\alpha}$ we have

$$A(\hat{\alpha}) = \int_0^{\hat{\alpha}} \phi''(t)\mathrm{d}t = \phi'(\hat{\alpha}) - \phi'(0) = -\phi'(0). \qquad (3.6)$$

Equation (3.6) allows an interesting perspective: although we do not know $\hat{\alpha}$, we do know $A(\hat{\alpha})$, the area above ϕ'' from 0 to $\hat{\alpha}$: it is $-\phi'(0)$.

Algorithms differ in the way they estimate $A(\alpha)$. Let $\hat{A}(\alpha)$ denote the estimate of $A(\alpha)$. Then many well-known algorithms can be reproduced by equating $\hat{A}(\alpha)$ with the true area $A(\hat{\alpha}) = -\phi'(0)$ and solving it for α. The estimating equation is given by

---

*In fact, it can be any search direction, but we have not considered any others thus far.

$$\hat{A}(\alpha) = A(\hat{\alpha}) = -\phi'(0) \qquad (3.7)$$

As an example consider the Newton–Raphson algorithm. It uses the rectangular estimate $\hat{A}_{NR}(\alpha) = \alpha\phi''(0)$, namely, the box with height $\phi''(0)$ and length α. Compare this in Figure 3.3. Equation (3.7) then gives the usual Newton–Raphson iterate. It is now possible to give a simple condition for the step-length to be monotonic. It connects the idea of *area overestimation* with the *monotonicity of an algorithm.*

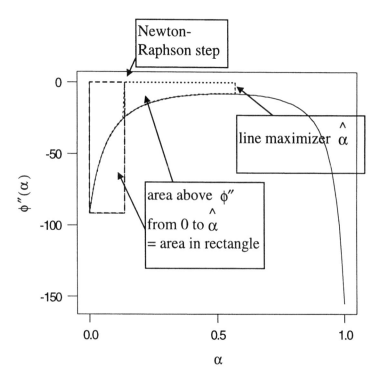

Figure 3.3. Illustration of the area estimator $A_{NR}(\alpha) = \alpha\phi''(0)$

Theorem 3.3: Let

$$\hat{A}(\alpha) \le A(\alpha) \text{ for all } \alpha \in [0, 1]. \qquad (3.8)$$

Then α^* defined by the estimating equation (3.7) is monotonic.

Proof. We have $\hat{A}(\alpha) \leq A(\alpha)$ for all α. In particular, this is true for α^*. Then $A(\alpha^*) \geq \hat{A}(\alpha^*) = A(\hat{\alpha})$. Since A is strictly decreasing we have $0 \leq \alpha^* \leq \hat{\alpha}$, implying $\phi(0) \leq \phi(\alpha^*)$. This ends the proof.

Note that $A(\alpha) \leq 0$. Geometrically, the condition (3.8) means that the area above the curve is always smaller than the estimated area.

In the case of $\hat{A}_{NR}$ this condition is met if $\min_{t \in [0, 1]} \phi''(t) = \phi''(0)$. In general, this is no gain since one optimization problem is replaced by another one. However, as we have noted above, ϕ'' is concave, and therefore, the minimum of ϕ'' is attained at one of the end points of the interval $[0, 1]$: $\min_{t \in [0, 1]} \phi''(t) = \min \{\phi''(0), \phi''(1)\}$.

Example 3.4: It is easy to construct a monotonic version of the Newton–Raphson step. Consider $\hat{A}_{box}(\alpha) = \alpha \min\{\phi''(0), \phi''(1)\}$. Then, $\hat{\alpha}_{box} = -\phi'(0)/\min\{\phi''(0), \phi''(1)\}$ is a monotonic step-length. For the construction of other monotonic step-length estimators, see Böhning (1989) or Böhning and Lindsay (1988).

In the situation of Figure 3.3 the minimum is attained at 1, though the ordinary Newton–Raphson step would still be monotonic. Thus, the danger with using $\hat{\alpha}_{box}$ lies in the fact that it might be too *conservative* and thus, slow down the convergence process. Let us investigate the question: what in Figure 3.3 determines the monotonicity of the ordinary Newton-Raphson step? Apparently, the area above ϕ'' from 0 to 1 is smaller than the rectangle area $1 \times \phi''(0) = \max\{\phi''(0), \phi''(1)\}$. This is summarized in the following theorem.

Theorem 3.4: Let $\phi''(0) \leq A(1) = \phi'(1) - \phi'(0)$. Then $\hat{\alpha}_{NR}$ is monotonic.

Proof. Let us assume without limitation of generality that $\max \{\phi''(0), \phi''(1)\} = \phi''(0)$. Suppose

$$\hat{A}_{NR}(\alpha) = \alpha\phi''(0) > A(\alpha) \text{ for some } \alpha \in (0, 1] \qquad (3.9)$$

Then, also $\phi''(\alpha) < \phi''(0)$. Since ϕ'' is concave, it is decreasing for $\alpha \geq \alpha_{max}$, where the latter is the maximizing value of ϕ''. Thus, $\phi''(t) \leq \phi''(\alpha) \leq \phi''(\alpha_{max})$ for all $t \in [\alpha, 1]$, Now, $\int_\alpha^1 \phi''(t)dt \leq (1-\alpha)\phi''(0)$, implying

$$\int_\alpha^1 \phi''(t)dt + \int_0^\alpha \phi''(t)dt < (1-\alpha)\phi''(0) + \alpha\phi''(0) = \phi''(0),$$

or

$$\int_0^1 \phi''(t)\mathrm{d}t = A(1) < \phi''(0)$$

which contradicts our condition stated in the theorem $\phi''(0) \le A(1)$. There-
fore the assumption (3.10) is false, and for all $\alpha \in [0, 1]$ $\hat{A}_{\mathrm{NR}}(\alpha) \le A(\alpha)$.
Thus, the condition (3.9) is met and the result follows with Theorem 3.3.
This ends the proof.

Let us now investigate the case that the condition of Theorem 3.3
is not met, in other words, $\phi''(0) > A(1)$. Then there exists a convex
combination $(1 - \alpha)\phi''(0) + \alpha\phi''(1)$ equalizing $A(1)$. Obviously, this
convex combination is given through $\{A(1) - \phi''(0)\}/\{\phi''(1) - \phi''(0)\}$.
Therefore the area estimate $\hat{A}_{\mathrm{sec}}(\alpha) = \alpha A(1) = \alpha[\phi'(1) - \phi'(0)]$ fulfills
condition (3.8). Thus, the solution of the estimating Equation (3.7)
$\hat{\alpha}_{\mathrm{sec}} = -\phi'(0)/[\phi'(1) - \phi'(0)]$ is monotonic. We summarize this result in
the following

Theorem 3.5: Let $\max\{\phi''(1), \phi''(0)\} > A(1)$. Then $\hat{\alpha}_{\mathrm{sec}}$ is monotonic.

Note the $\hat{\alpha}_{\mathrm{sec}}$ is the well-known *secant* method. All these different
step-lengths can be chosen in a submenu of C.A.MAN which is dis-
cussed in the next chapter.

3.4   C.A.MAN (computer-assisted analysis of mixtures)

For the practical realization of these algorithms a progamming package has been developed by the name of C.A.MAN (Computer-assisted
analysis of mixtures).* In this section we focus on the way in which
C.A.MAN works. We recall that the goal is to maximize $l(P)$ in the
simplex Ω of all probability distributions P on parameter space λ.
We call the solution of this problem the fully iterated nonparametric
maximum likelihood estimator $\hat{P}$ and this solution is achieved in
C.A.MAN in two phases.

In phase I, an approximating grid $\lambda_1, ..., \lambda_L$ $(L \le 50)$ is chosen and
$l(P)$ is maximized in the simplex Ω_{grid} of all probability distributions
P on grid $\{\lambda_1, ..., \lambda_L\}$ with one of the algorithms described above. For

---

*C.A.MAN can be downloaded from the homepage of the author:
www.medizin.fu-berlin.de/sozmed/bo1.html

example, we can choose the observed data values as an approximating grid, in the introductory example $L = 21$ different values of sold number of packages were observed: $\{\lambda_1, ..., \lambda_L\} = \{0, 1, 2, ..., 20\}$ with $L = 21$. As potential choice of initial weights the observed relative frequencies or uniform weights $p_i = 1/21$ could be used.

In phase II, all grid points which are left with positive weights as a result of the optimization process in phase I are used as initial values for the EM algorithm (Dempster, Laird, and Rubin 1977) — a concept to be discussed in the next section — to produce the fully iterated NPMLE.

Example 3.5: The discussion from Example 1.2 is continued and is used to illustrate some results of the analysis with C.A.MAN. In *Phase I* the options VEM with optimal step-length are used.

The grid consists of all observed data values. i.e., 0, 1, 2, ..., 20. Table 3.1 shows the results after the stopping rule 3.1 with $\varepsilon = 0.000001$ ($\max_{\lambda \in \Omega_{grid}} d(\lambda, P_n) < 1 + \varepsilon$) has been achieved. C.A.MAN could identify 7 points of support and the log-likelihood at iterate -1130.61200. See Table 3.2.

λ_j	p_j	$d(\lambda_j, P_n)$	λ_j	p_j	$d(\lambda_j, P_n)$
0	0.1554	1.	11	0.	0.9983
1	0.1272	1.	12	0.	0.9996
2	0.	0.9977	13	0.0968	1.
3	0.4026	1.	14	0.	0.9951
4	0.0725	1.	15	0.	0.9798
5	0.	0.9993	16	0.	0.9491
6	0.	0.9995	17	0.	0.9003
7	0.0540	1.	18	0.	0.8337
8	0.0914	1.	19	0.	0.7520
9	0.	0.9991	20	0.	0.6601
10	0.	0.9982			

Table 3.1. Results of VEM algorithm after 1560 iterations

λ_j	p_j	$d(\lambda_j, P_n)$	λ_j	p_j	$d(\lambda_j, P_n)$
0	0.1554	1.	7	0.0540	1.
1	0.1272	1.	8	0.0914	1.
3	0.4026	1.	13	0.0968	1.
4	0.0725	1.			

Table 3.2. Identification of support points

In *Phase II*, these 7 parameter values are used as *initial values* of the
EM algorithm, a procedure discussed in the next section, but we present
the results here.

λ_j	p_j	λ_j	p_j
0	0.0002	7.418164	0.0563
0.204903	0.2440	7.418164	0.0951
3.001951	0.4314	12.872540	0.1017
3.001951	0.0713		

Table 3.3. Results after 3689 iterations of the EM algorithm

Obviously, there are two cases with *identical* components, namely,
components 3 and 4 as well as components 5 and 6. These are collapsed
in Table 3.4, which then represents the fully iterated maximum likelihood estimator.

λ_j	p_j	λ_j	p_j
0	0.0002	7.418164	0.1514
0.204903	0.2440	12.872540	0.1017
3.001951	0.5027		

Table 3.4. Results after collapsing identical values

The log-likelihood at $\hat{P}$ is -1130.07100. In comparison, the log-likelihood
at iterate based on the approximating grid is $l(\hat{P}_{grid}) = -1130.61200$. This
was the state investigators had reached when the C.A.MAN approach
was first published in 1992. In the sequel, C.A.MAN was often criticized
because it is not really computing the NPMLE, but only an approximating solution. This example (as do many others) shows that the gain to
the fully iterated solution is *not* great and the solution provided by the
approximating approach would have been satisfactory. Figure 3.4 shows
that the iterated solution is indeed the NPMLE.

Packages for mixture distribution modeling have been recently
reviewed in *The American Statistician* (Haughton 1997). In this review,
in addition to C.A.MAN four other packages are considered: BINOMIX
for binomial mixtures (Erdfelder 1993), MIX by MacDonald (1986) and
MacDonald and Pitcher (1979), and the routines in the larger packages
BMDP and STATA. With regard to C.A.MAN, Haughton (1997) writes
that the program is menu driven and very easy to use, and further
that C.A.MAN is more robust with respect to initial values* (*in*

*Since the algorithms developed above are globally covergent.

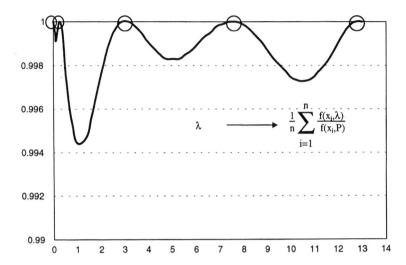

Figure 3.4. Gradient function at fully iterated NPMLE for Example 1.2.

comparison to the other packages, D.B.) and offers an interesting semi-parametric estimation option (Haughton 1997, p. 204). One should mention that there are other routines for computing mixture distribution estimates available, including DerSimonian (1986, 1990), Agha and Ibrahim (1984), or McLachlan (1996).

3.5 The EM algorithm for the fixed component case

Let for the time being $\lambda_1, \lambda_2, \ldots, \lambda_k$ be *fixed* and *known*, and denote f(x, $\boldsymbol{p}$) = p_1 f(x, λ_1) + p_2 f(x, λ_2) + ... + p_k f(x, λ_k) as the mixture density. Because $p_1 + p_2 + \ldots + p_k = 1$, we can write f(x, $\boldsymbol{p}$) = p_1f(x, λ_1) + p_2f(x, λ_2) + ... + p_{k-1}f(x, λ_{k-1}) + $(1 - \Sigma_i p_i)$f(x, λ_k) with $\boldsymbol{p} = (p_1, p_2, \ldots, p_{k-1})^\mathrm{T}$ and $\Sigma_i \, p_i = p_1 + p_2 + \ldots + p_{k-1}$.

Estimation of the parameters $p_1, \ldots, p_{k-1}$ by maximizing the log-likelihood function leads to

$$l(p_1, \ldots, p_{k-1}) = \Sigma_x \log[\Sigma_{j=1}^{k-1} f(x, \lambda_j)p_j + (1 - \Sigma_{j=1}^{k-1} p_j)f(x, \lambda_k)] .$$

Taking partial derivatives with respect to p_j gives

$$\frac{\partial l}{\partial p_j} = \Sigma_x \frac{f(x, \lambda_j) - f(x, \lambda_k)}{f(x, \boldsymbol{p})} \text{ for } j = 1, 2, \ldots, k-1$$

and leads to the likelihood equations

$$\frac{\partial l}{\partial p_j} = \Sigma_x \frac{f(x, \lambda_j) - f(x, \lambda_k)}{f(x, p)} = 0, j = 1, 2, ..., k-1. \qquad (3.10)$$

Since $\Sigma_x f(x, \lambda_j)/f(x, p) = \Sigma_x f(x, \lambda_k)/f(x, p)$ for $j = 1, ..., k-1$, it follows that

$$\Sigma_{j=1}^{k-1} p_j \Sigma_x f(x, \lambda_j)/f(x, p) = \Sigma_{j=1}^{k-1} p_j \Sigma_x f(x, \lambda_k)/f(x, p)$$

and adding $\Sigma_x p_k f(x, \lambda_k)/f(x, p)$ on both sides leads to

$$\Sigma_x f(x, \lambda_k)/f(x, p) = \Sigma_x f(x, p)/f(x, p) = n \qquad (3.11)$$

and the likelihood equations take on the form

$$\Sigma_x f(x, \lambda_j)/f(x, p) = n \text{ for } j = 1, ..., k. \qquad (3.12)$$

Or yet another form is that likelihood equations turn into the *fixed point* equation

$$\Sigma_x \frac{p_j f(x, \lambda_j)}{n f(x, p)} = p_j \text{ for } j = 1, ..., k. \qquad (3.13)$$

Equation (3.13) can be written more compactly as $F(p) = p$, where the jth component $F_j(p)$ is given by the left-hand side of (3.13). An example of F_1 is given in Figure 3.5 for $k = 2$. As it will turn out, (3.13) is a special case of a more general fixed point algorithm, the EM algorithm, an algorithmic concept developed by Dempster, Laird, and Rubin (1977). For a detailed investigation of the convergence properties, see Wu (1983) or McLachlan and Krishnan (1997). We will now give an outline of the EM algorithm in the mixture model context. The EM algorithm requires two log-likelihoods: the one we have observed is called the *incomplete* log-likelihood and is given for the mixture model as

$$l(p_1, ..., p_{k-1}) = \Sigma_x \log[\Sigma_{j=1}^{k-1} f(x, \lambda_j) p_j + (1 - \Sigma_{j=1}^{k-1} p_j) f(x, \lambda_k)].$$

The other one enters the setting through a *latent, unobserved* variable Z_j, which takes on the value 1 if x is from sub-population j, and 0 otherwise. Let z_{ij} denote the value of Z_j for observation x_i. If z_{ij} were observable we could estimate p_j simply by the number of times z_{ij} is

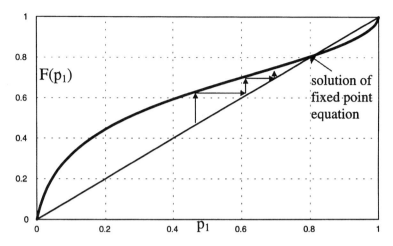

Figure 3.5. Fixed point mapping (3.14) for $k = 2$.

equal to 1, for given j, divided by n. Thus, $\hat{p}_j = \Sigma_i\, z_{ij}/n$, for $j = 1, ..., k$. The likelihood for the pair $(x_i, z_{ij})^\mathrm{T}$ is

$$\Pr(X_i = x_i, Z_{i1} = z_{i1}, ..., Z_{ik} = z_{ik})$$

$$= \Pr(X_i = x_i | Z_{i1} = z_{i1}, ..., Z_{ik} = z_{ik}) \times \Pr(Z_{i1} = z_{i1}, ..., Z_{ik} = z_{ik})$$

$$= \prod_{j=1}^{k} f(x_i, \lambda_j)^{z_{ij}} p_j^{z_{ij}}$$

Note that the mixture or marginal likelihood and the complete likelihood are connected via

$$\sum_{z} \prod_{j=1}^{k} f(x, \lambda_j)^{z_j} p_j^{z_j} = \sum_{j=1}^{k} f(x, \lambda_j) p_j,$$

where the sum on the left-hand side is taken over all vectors $z = (z_1, ..., z_k)^\mathrm{T}$ having a single 1 at one position, whereas the rest are zeros (there are exactly k of those). For the full likelihood we have

$$\prod_{i=1}^{n} \prod_{j=1}^{k} f(x_i, \lambda_j)^{z_{ij}} p_j^{z_{ij}}.$$

This is called the complete likelihood and the associated complete log-likelihood is

$$l_{\text{com}}(\boldsymbol{p}) = \Sigma_i \Sigma_j z_{ij} \log(p_j) + C \tag{3.14}$$

where $C = \Sigma_i \Sigma_j z_{ij} \log \mathrm{f}(x_i, \lambda_j)$ is independent of $\boldsymbol{p}$. Note that the maximization of (3.14) leads to $\hat{p}_j = \Sigma_i z_{ij}/n$ for $j = 1, ..., k$. What remains is to clarify what to do with the z_{ij} since they have not been observed. It is natural to replace them with their expected values given the data $x_1, ..., x_n$ and $\boldsymbol{p}$: $E(z_{ij}|x_i, \boldsymbol{p}) = \mathrm{Pr}_{\boldsymbol{p}}(Z_{ij} = 1|X_i = x_i)$. Now, using Bayes' theorem, we find that

$$\mathrm{Pr}_{\boldsymbol{p}}(Z_{ij} = 1|X_i = x_i) \tag{3.15}$$

$$= \frac{\mathrm{Pr}_{\boldsymbol{p}}(X_i = x_i|Z_{ij} = 1)\mathrm{Pr}(Z_{ij} = 1)}{\Sigma_{j'}\mathrm{Pr}_{\boldsymbol{p}}(X_i = x_i|Z_{ij'} = 1)\mathrm{Pr}(Z_{ij'} = 1)}$$

$$= \frac{\mathrm{f}(x_i, \lambda_j)p_j}{\Sigma_{j'}\mathrm{f}(x_i, \lambda_{j'})p_{j'}} =: e_{ij}$$

Replacing z_{ij} by (3.15) leads to the expected complete log-likelihood

$$E\{l_{\text{com}}(\boldsymbol{p})\} = \Sigma_i \, \Sigma_j \, e_{ij} \log(p_j) + C. \tag{3.16}$$

This constitutes the *E-step* in the EM algorithm. Maximizing (3.16) leads to the *M-step* and the solution

$$p_j^{(\text{new})} = \Sigma_i e_{ij}/n = \Sigma_i \frac{\mathrm{f}(x_i, \lambda_j)p_i}{\Sigma_{j'}\mathrm{f}(x_i, \lambda_{j'})p_{j'}}/n = \Sigma_i \frac{\mathrm{f}(x_i, \lambda_j)p_i}{n\Sigma_{j'}\mathrm{f}(x_i, \lambda_{j'})p_{j'}} \tag{3.17}$$

In summary, the EM algorithm for fixed component mean values is given as

Algorithm 3.4 (EM algorithm for known component densities):

Step 0. Let $\boldsymbol{p}$ be any initial vector of weights.

Step 1. Compute the E-step according to (3.15).

Step 2. Compute the M-step according to (3.17) leading to $\boldsymbol{p}^{(\text{new})}$.

Step 3. Set $\boldsymbol{p} = \boldsymbol{p}^{(\text{new})}$ and go to step 1.*

Example 3.6: We demonstrate the EM algorithm in the following setting which has been already used in Böhning (1989). Suppose that there are 3 values of λ: λ_1, λ_2, λ_3 and $x \in \{1, 2, 3, 4\}$. Let

$$[f(x, \lambda_1)] = (0.60\ 0.30\ 0.05\ 0.05)^T$$

$$[f(x, \lambda_2)] = (0.05\ 0.15\ 0.30\ 0.50)^T$$

$$[f(x, \lambda_3)] = (0.01\ 0.08\ 0.21\ 0.70)^T$$

and x be observed 15 times for $x = 1$, 10 times for $x = 2$, 20 times for $x = 3$, and 55 times for $x = 4$. Thus, $n = 100$.

It becomes quite evident in Table 3.5 that the convergence behavior is rather slow. After 100 iterations we have only *one* correct decimal place, after 1000 about 2 to 3, and only after 10,000 iterations does the convergence become satisfactory.

Iteration	p_1	p_2	p_3
1	1/3	1/3	1/3
10	0.1804	0.3272	0.4966
100	0.2043	0.0994	0.6968
1000	0.2103	0.0426	0.7471
10000	0.2102	0.0424	0.7473
∞	0.2102	0.0424	0.7473

Table 3.5. Results of the EM algorithm in Example 3.6

Now, suppose that the component parameters are *unknown* themselves. This means that we have to take into account the expression $C = \Sigma_i \Sigma_j e_{ij} \log f(x_i, \lambda_j)$ in (3.16). Since the total log-likelihood separates into two parts, the first depending on p only, the second depending on $\lambda_1, \ldots, \lambda_k$ only, we can incorporate estimation of $\lambda_1, \ldots, \lambda_k$ by finding the maximum of $\Sigma_i \Sigma_j e_{ij} \log f(x_i, \lambda_j)$. This will depend on the specific form of $f(x, \lambda)$. Let, for example $f(x, \lambda)$ be the Poisson density (2.4): $f(x_i, \lambda, E_i) = \exp(-\lambda E_i) (\lambda E_i)^{x_i}/x_i!$. Then maximization of C leads to

$$\lambda_j^{(new)} = \frac{\Sigma_i e_{ij} x_i}{\Sigma_i e_{ij} E_i} \qquad (3.18)$$

*It should be noted that a suitable stopping rule should be incorporated.

Iteration	p_1	p_2	λ_1	λ_2
1	1/2	1/2	1	2
10	0.7002	0.2998	0.0989	0.4463
100	0.6853	0.3147	0.0763	0.4791
1000	0.5729	0.4271	0.0441	0.4163
10000	0.4779	0.5251	0.0124	0.3755
∞	0.4779	0.5251	0.0124	0.7473

Table 3.6. Results of the EM algorithm for $f(x_i, \lambda) = \exp(-\lambda)(\lambda)^{x_i}/x_i!$ and data discussed in Example 3.7

Example 3.7: In traffic accident research, Kuan *et al.* (1991) discuss data coming from the California Department of Motor Vehicles master driver license file. Here the variable of interest is the number of accidents per driver which range from 0 to 3. A possible motivation can be seen in the possibility of finding risk factors involved in the accidents. There are 5422 drivers in the file with 4499 having 0 accidents, 766 having 1 accident, 136 with 2 accidents and 21 with 3 accidents. It has been suggested to consider a 2-component Poisson mixture model for this data set.

Consider as a further example the exponential density, $f(x, \lambda) = (1/\lambda)e^{-x/\lambda}$. Here, maximization of C leads to

$$\lambda_j^{(new)} = \Sigma_i e_{ij} x_i / \Sigma_i e_{ij}. \tag{3.19}$$

Furthermore, consider the binomial $f(x, \lambda, N) = \binom{N}{x}\lambda^x(1-\lambda)^{N-x}$. We find for the M-step

$$\lambda_j^{(new)} = \Sigma_i e_{ij} x_i / \Sigma_i e_{ij} N_i. \tag{3.20}$$

Evidently, in all situations considered thus far the M-step leads to a form of *weighted* mean of the observations. As a slightly different example, we consider the *geometric* distribution $f(x, \lambda) = (1-\lambda)^x \lambda$ for $x = 0, 1, 2, \ldots$ and $\lambda \in [0, 1]$. Here we find that

$$\lambda_j^{(new)} = \frac{\Sigma_i e_{ij}}{\Sigma_i e_{ij} x_i + \Sigma_i e_{ij}}. \tag{3.21}$$

Let us finally consider the normal density

$$f(x, \lambda_j) \;=\; \frac{1}{\sqrt{2\pi\sigma^2}} \exp\left\{-\frac{1}{2}(x - \lambda_j)^2/\sigma^2\right\}.$$

Again, we find that $\lambda_j^{(\text{new})}$ is provided by (3.19). However, we have also to consider the additional parameter σ^2. Taking the derivative of C with respect to σ^2 leads to the likelihood equation

$$\Sigma_i\, \Sigma_j\, e_{ij}(x_i - \lambda_j)^2/\sigma^4 - \Sigma_i\, \Sigma_j\, e_{ij}/\sigma^2 = 0$$

and the solution

$$\sigma^2_{(\text{new})} \;=\; \Sigma_i \Sigma_j e_{ij}(x_i - \lambda_j)^2 / \Sigma_i \Sigma_j e_{ij}\,.$$

Since $\Sigma_j\, e_{ij} = 1$ for all i (for fixed i the e_{i1}, ..., e_{ik} establish a discrete probability distribution), this solution is simply

$$\sigma^2_{(\text{new})} \;=\; \frac{1}{n}\Sigma_i\Sigma_j e_{ij}(x_i - \lambda_j)^2 \qquad\qquad (3.22)$$

Algorithm 3.5 (EM algorithm for mixture models):

Step 0. Let $\boldsymbol{p}$ be any initial vector of weights and λ any initial vector of component parameters.

Step 1. Compute the E-step according to (3.15).

Step 2. Compute the M-step according to (3.17) leading to $\boldsymbol{p}^{(\text{new})}$ and the appropriate update form (3.18)–(3.21) leading to $\lambda^{(\text{new})}$. In the case of the normal distribution the variance parameter is updated according to (3.22).

Step 3. Set $\boldsymbol{p} = \boldsymbol{p}^{(\text{new})}$, and $\lambda = \lambda^{(\text{new})}$ and go to step 1.

It is one of its famous properties that the EM iteration does not decrease the log-likelihood and therefore guarantees convergence (see McLachlan and Krishnan 1997 for a detailed theoretical investigation of convergence). However, this property does not imply convergence to a global maximum, as the following example demonstrates.

Example 3.8: We consider an application in nutritional anthropometry. For the evaluation of the nutritional status of children in developing countries, the anthropometric standardized scores WE/HE, WE/AGE, and HE/AGE based on an international reference population are recommended (Waterlow *et al.*, 1977, Böhning and Schelp 1986). For example, HE/AGE is computed as follows: given the variables SEX, HEIGHT, and AGE of an individual child, height is evaluated in terms of a reference population (NCHS, 1976), which contains various statistical measures (based on large numbers of observations) including median height and standard deviation for a given age and sex class. Then, the nutritional indicator HE/AGE is computed as (HEIGHT − M)/SD, where HEIGHT is the child's height and M is the median height of the reference population according to the age and sex class of the child. These indicators are normalized to have variance and mean 0; however, the latter only if *no* malnourishment occurs. For the detection of subclinical malnourishment, we could consider a mixture model consisting of two normal distributions. For a sample of 520 preschool children from NE Thailand and their associated HE/AGE-scores, we use C.A.MAN and the EM algorithm to find maximum likelihood estimates for the four parameters of the 2-component normal mixture model. Table 3.7 shows the results of the EM algorithm for three sets of initial values. Evidently, the first set yields the highest likelihood.

Setting	p_1	λ_1	λ_2	l	σ^2
initial values	0.9	0	−7		1.
at convergence	0.9907	−1.6194	−5.8535	−680.6718	0.7305
initial values	0.5	0	6		1.
at convergence	0.9962	−1.6472	6.8700	−695.1904	0.8270
initial values	0.5	−0.5	0.5		1.
at convergence	0.8355	−1.6749	−1.5034	−687.5996	0.8232

Table 3.7. Results of the EM algorithm in Example 3.8

3.6   The choice of initial values for the EM algorithm

The EM algorithm (Dempster *et al.* 1977) has become a popular algorithm for finding maximum likelihood estimates (McLachlan and Krishnan 1997). Recently, attention has focussed on improving its convergence behavior or finding faster converging alternatives (Jamshidian and Jennrich 1997; Kowalski *et al.* 1997; Meng and van Dyk 1997). The EM algorithm — as with any algorithm — has to be started with certain values for the parameters it is optimizing. Usually, the literature is vague on this question, or does not mention it at all.

Usually, the problem is treated in an ad hoc manner, leaving the reader with the impression that the choice of starting values is not too crucial.

In fact, it is, or, to be more careful, it can be. In this section we discuss a situation where it turns out that the choice of initial values is the most important problem. Working with the conventional EM algorithm (in contrast to accelerated versions) means only that we are using more computer time, but the answers are right or wrong depending on the right or wrong initial values.

Often the EM algorithm is used in examples with a structure supporting the model under investigation. In these cases, it is admitted that the choice of starting values might indeed be less crucial. Consider, however, the following situation.

Let $f(x, \lambda)$ be a parametric density and $f(x, \lambda, p) = \sum_{j=1}^{k} f(x, \lambda_j) p_j$ be a mixture of these ($p_j \geq 0$, $p_1 + \dots + p_k = 1$). We are interested to test H_0: $k = 1$ against H_1: $k = 2$. Given a sample $x_1, \dots, x_n$, the likelihood ratio statistic

$$2\log\xi = 2\Sigma_i\left[\log\left(\sum_{j=1}^{2} f(x_i, \hat{\lambda}_j)\hat{p}_j\right) - \log f(x_i, \bar{x})\right]$$

$$= 2\Sigma_i[\log(\hat{p}f(x_i, \hat{\lambda}_1) + (1 - \hat{p})f(x_i, \hat{\lambda}_2)) - \log f(x_i, \bar{x})]$$

is not standard under H_0 and percentiles of it have been computed for various densities of the exponential family using simulation techniques (Böhning *et al.* 1994; see also Chapter 4). In these cases, using the EM algorithm the maximum likelihood estimates $\hat{\lambda}_1$, $\hat{\lambda}_2$, $\hat{p}_1 = \hat{p}$, $\hat{p}_2 = 1 - \hat{p}$ have to be computed under a model not supported by the data. As it turns out, at least in these instances the choice of initial values has to be given the highest attention and by no means can be treated in an ad hoc manner. Here, various strategies for choosing initial values are investigated and evaluated.

Strategies for Choosing Initial Values. Three methods for choosing initial values are considered. In the first two strategies starting values for λ_1, λ_2, and p are found by classifying the n data points into *two disjoint* sets of size m and $n - m$, respectively ($m < n$). Then λ_1 is estimated by the arithmetic mean of the first set and λ_2 by the arithmetic mean of the second set. p is estimated as m/n. In particular, ordered values $x_{(1)}, \dots, x_{(n)}$ are considered, and the sets are updated by including one observation at each step. Thus, at first, set 1 consists of $x_{(1)}$, set 2 of $x_{(2)}, \dots, x_{(n)}$. In the second step, set 1 consists of $x_{(1)}, x_{(2)}$ whereas set 2 consists of $x_{(3)}, \dots, x_{(n)}$, and so forth. The procedure considers $n - 1$ partitions. Strategies differ in the way they select the optimal partition.

Strategy I maximizes

$$l_1(\lambda_1, \lambda_2, p) = \Sigma_i \log(p\ f(x_i, \lambda_1) + (1 - p)\ f(x_i, \lambda_2))$$

in $\lambda_1 = \bar{x}_1$, $\lambda_2 = \bar{x}_2$, and $p = m/n$, meaning that the log-likelihood has to be evaluated $n - 1$ times.

Strategy II is minimizing the total sum of squares in m:

$$l_{II} = \sum_{i=1}^{m} (x_{(i)} - \bar{x}_1)^2 + \sum_{i=m+1}^{n} (x_{(i)} - \bar{x}_2)^2$$

where $\bar{x}_1$ and $\bar{x}_2$ are the means of the first m and the remaining $n - m$ ordered values, respectively. This strategy is attractive, because it can be implemented in a faster way by maximizing

$$\left(\sum_{i=1}^{m} r_{(i)} \right)^2 / m + \left(n\bar{r} - \sum_{i=1}^{m} r_{(i)} \right)^2 / (n - m)$$

in m. This requires only $2n$ flops.[*]

In Strategy III the means are chosen as values of certain order statistics. The following values were chosen: $\lambda_1 = x_{(1)}$, $\lambda_2 = x_{(n)}$ (Strategy III.1), $\lambda_1 = x_{(5)}$, $\lambda_2 = x_{(n-5)}$ (Strategy III.5), $\lambda_1 = x_{(30)}$, $\lambda_2 = x_{(n-30)}$ (Strategy III.30). p is chosen to be $1/2$ in all three cases of Strategy III. Strategy III.1 has been recommended in Böhning et al. (1992). In Böhning et al. (1994) — for the exponential distribution — starting values according to $\lambda_{168,2} = \lambda \pm 1/2\lambda$ were used, where λ is the population parameter. Let $n = 100$. Since the standard error of $\bar{x}$ is $\lambda/\sqrt{n} = \lambda/10$, starting values according to $\lambda \pm 1/2\lambda$ imply that we are starting at $\pm 5 \times$ S.E., being well in the extremes of the distribution. However, as mentioned by Seidel et al. (1997, 1999) this strategy turned out to be *not* favourable even for larger sample sizes.

Design of the Simulation Study. Samples of size $n = 100$ were generated from an exponential distribution ($f(x, \lambda) = (1/\lambda) \exp\{-x/\lambda\}$), where λ takes values 1, 2, 3, 4, 5. For each of these 5 populations the samples were replicated $B = 100$ times. For each of these replications the likelihood ratio $2 \log \xi = 2[l(\hat{\lambda}_1, \hat{\lambda}_2, \hat{p}) - (-n\log(\bar{x}) - n)]$ was calculated. Here, $-n\log(\bar{x}) - n$ is the log-likelihood under the null

[*]*flop* is the abbreviation of floating point operation used in computational analysis.

hypothesis H_0: $k = 1$, and is independent of any initial value. $l(\lambda_1, \lambda_2, \hat{p})$ is the log-likelihood under the alternative H_1: $k = 2$, and it depends on the initial values. For each of the 5 strategies initial values had to be computed, the EM algorithm had to be started and limiting points had to be iterated. These were used to compute the log-likelihood which was then applied in the likelihood ratio test.

Although the 5 methods could be compared against each other, it might be the case that none of the methods is achieving the global maximum in all cases. Therefore, it was sought to establish a gold standard. This was accomplished by placing a fine grid on the data space and starting the EM algorithm from all the grid points. In particular, λ_1 varied from $x_{(1)}$, $x_{(5)}$, ..., $x_{(95)}$ and, for each value of λ_1, λ_2, varied from $x_{(j+1)}$, ..., $x_{(100)}$ where $x_{(j)}$ is defined through $\lambda_1 = x_{(j)}$. p is chosen to be $^1\!/_2$, since the choice of the weight is less crucial in mixture models. In fact, conditional on the values of λ_1 and λ_2, there is always a unique global maximizing weight since the log-likelihood is easily seen to be strictly concave in p (for $\lambda_1 \neq \lambda_2$).

Results. The results are summarized in Table 3.8 for the 95th percentile of the likelihood ratio distribution and in Table 3.9 for the 99th percentile. Strategy II fails to reach the global maximum in many cases, and will therefore produce a strong underestimation of the desired percentiles. This is similarly true for Strategy III, when the initial values are chosen as the 30th and 70th order statistic (Strategy III.30). It is remarkable to see that it even makes a difference if the initial values

λ	Gold	I	II	III.1	III.5	III.30
1	5.98950	5.98950	3.01275	5.98950	5.95936	5.36394
2	4.13227	4.13227	2.67773	3.97518	3.80745	2.67756
3	5.21104	5.21080	3.40843	5.21104	5.10187	3.44138
4	6.42586	6.31037	4.21047	6.30965	6.42581	4.30117
5	4.13227	4.13227	2.78511	4.10180	3.46699	2.77548

Table 3.8. Achieved values for 95th percentile for 2 log ξ

λ	Gold	I	II	III.1	III.5	III.30
1	9.00509	9.00502	6.71142	9.00507	9.00502	6.71072
2	9.21600	9.21600	3.84848	9.21600	5.46918	3.80979
3	8.02484	8.02484	5.71976	8.02498	8.02491	5.71933
4	9.00035	9.00035	8.75463	9.00030	9.00021	8.75392
5	12.52232	12.52232	6.48851	12.52227	6.48627	6.48804

Table 3.9. Achieved values for 99th percentile for 2 log ξ

are chosen according to minimum and maximum of the data (Strategy III.1) or according to the 5th and 95th order statistic. Strategy I (maximizing the classification likelihood) is expensive, but good.

CHAPTER 4

Likelihood ratio test and number of components

4.1 Diagnostic (overdispersion) tests for heterogeneity

In this chapter we look at testing methods which provide some basis for deciding about the presence and structure of heterogeneity. We begin with the situation that we are interested in deciding about the presence or absence of heterogeneity. This might be formulated as the hypothesis H_0: $k = 1$ (*homogeneity*) versus H_1: $k > 1$. Note that we are *not* formulating some particular alternative such as $k = k^*$. For testing this hypothesis we can employ a simple diagnostic test, commonly known as an *overdispersion test*.

The presence of population heterogeneity and the occurrence of overdispersion are connected as follows. Suppose we are interested in a random quantity X in a population consisting of k subpopulations with means λ_j, variances σ_j^2, and weights p_j, $j = 1, ..., k$. Suppose further that this quantity is sampled under *ignorance* of the subpopulation structure. Then $E(X) = p_1\lambda_1 + ... + p_k\lambda_k = \bar{\lambda}$ and

$$\text{Var}(X) = \sum_{j=1}^{k} p_j\sigma_j^2 + \sum_{j=1}^{k} p_j(\lambda_j - \bar{\lambda})^2. \tag{4.1}$$

This relationship is derived in detail in Chapter 6. Now, suppose that we are in the one-parameter exponential family, such as the Poisson, binomial or exponential distribution. Then there exists a functional relationship between σ_j^2 and the mean λ_j such as $\sigma_j^2 = \lambda_j(1 - \lambda_j)n$ for the binomial with sample size parameter n, $\sigma_j^2 = \lambda_j$ for the Poisson, and $\sigma_j^2 = \lambda_j^2$ for the exponential. In the case of population *homogeneity* we simply have $Var(X) = \sigma^2(\lambda)$, where $\sigma^2(\lambda)$ is expressing the variance-mean relationship. Then (4.1) takes on the form

$$Var(X) = \sum_{j=1}^{k} p_j \sigma^2(\lambda_j) + \sum_{j=1}^{k} p_j (\lambda_j - \bar{\lambda})^2 = \sum_{j=1}^{k} p_j \sigma^2(\lambda_j) + \tau^2. \qquad (4.2)$$

where $\tau^2 = \sum_{j=1}^{k} p_j (\lambda_j - \bar{\lambda})^2$ is called the *heterogeneity* variance and is expressing the amount of extra-dispersion which is due to the population heterogeneity. Clearly, if there is homogeneity the variance-mean relationship holds, and if there is heterogeneity *overdispersion* occurs. See also Aitkin *et al.* (1990) or Hinde and Demétrio (1998) for a more general discussion. We exemplify these issues in the example of the Poisson; since $\sigma_j^2 = \lambda_j$ (4.2) takes on the form

$$Var(X) = \sum_{j=1}^{k} p_j \sigma^2(\lambda_j) + \tau^2 = E(X) + \tau^2 \qquad (4.3)$$

which is particularly simple. *Overdispersion tests* focus on the decomposition (4.2) in that they test the hypothesis $\tau^2 = 0$ without explicit knowledge of the form of heterogeneity. For example, we find for (4.3) the simple overdispersion estimate $\hat{\tau}^2 = S^2 - \bar{X}$, where $S^2 = 1/(n-1) \sum_{i=1}^{n} (X_i - \bar{X})^2$. As a simple feature of these procedures, it is possible to derive tests for the presence of heterogeneity without invoking, for example, maximum likelihood procedures with their inherent numerical complexity.

Definition 4.1 Any test of the hypothesis H_0: $\tau^2 = 0$ against H_1: $\tau^2 > 0$ is called an *overdispersion test*.

In the following we will consider a test in a special situation, namely, that we have a random sample of n counts, X_1, X_2, ..., X_n, and we are interested in finding out about the appropriateness of a *homogeneous* Poisson distribution. Tiago de Oliveira (1965) derived the following test statistic:

$$O_T = \sqrt{n} \frac{S^2 - \bar{X}}{\sqrt{1 - 2\sqrt{\bar{X}} + 3\bar{X}}} \qquad (4.4)$$

and claimed that (4.4) were approximately standard normal if the sample were from a homogeneous Poisson population with parameter λ. The test is referenced in Titterington, Smith and Makov (1985, p. 152), Johnson, Kotz, and Kemp (1992, p. 319), or Everitt and Hand (1981, p. 118), to mention just three excellent reference books. However, the test is using a *wrong* estimate of the variance of $S^2 - \overline{X}$ as is pointed out in Böhning (1994). In fact, if X is a Poisson variate with mean λ, then

$$(X - \lambda)^2 - X = (X - \lambda)^2 - (X - \lambda) - \lambda,$$

so that

$$
\begin{aligned}
\mathrm{Var}\{(X-\lambda)^2 - X\} &= \mathrm{Var}\{(X-\lambda)^2 - (X-\lambda) - \lambda\} \qquad (4.5)\\
&= E\{(X-\lambda)^2 - (X-\lambda)\}^2 - \lambda^2\\
&= E(X-\lambda)^4 - 2E(X-\lambda)^3 + E(X-\lambda)^2 - \lambda^2\\
&= \lambda + 3\lambda^2 - 2\lambda + \lambda - \lambda^2 = 2\lambda^2,
\end{aligned}
$$

where we have used that the third and fourth central moments of a Poisson variate are given by λ and $\lambda + 3\lambda^2$, respectively (see Haight 1967). It follows that

$$\mathrm{Var}\left\{\frac{1}{n}\sum_{i=1}^{n}[(X_i - \lambda)^2 - X_i]\right\} = \frac{1}{n}2\lambda^2.$$

Since for large n, $\mathrm{Var}\{1/n \sum_{i=1}^{n}[(X_i - \lambda)^2 - X_i]\}$ and $\mathrm{Var}(S^2 - \overline{X})$ should be close, the variance proposal $(1 - 2\sqrt{\lambda} + 3\lambda)/n$ given by Tiago de Oliveira cannot be correct. See Figure 4.1. The *exact* variance formula is given in the following theorem.

Theorem 4.1 (Böhning 1994): Let $X_1, ..., X_n$ be a sample from a Poisson distribution with parameter λ. Then

$$\mathrm{Var}(S^2 - \overline{X}) = 2\lambda^2/(n-1) \qquad (4.6)$$

Proof. We prove that $n^2(n-1)^2 E(S^2 - \overline{X})^2 = E[n \Sigma_i X_i^2 - \Sigma_i X_i (\Sigma_i X_i + n - 1)]^2 = 2n^2(n-1)\lambda^2$ and use at various points of the proof that the first four moments of a Poisson variate X are given by $EX = \lambda$, $EX^2 = \lambda + \lambda^2$, $EX^3 = \lambda + 3\lambda^2 + \lambda^3$, $EX^4 = \lambda + 7\lambda^2 + 6\lambda^3 + \lambda^4$ (Haight 1967, p. 6). We write

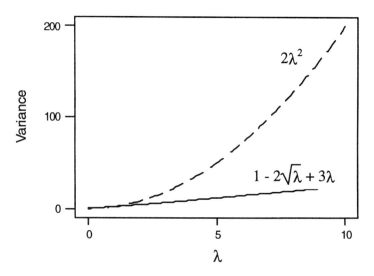

Figure 4.1. Two variance formulas for $S^2 - \overline{X}$ in comparison.

$n^9(n-1)^9(\mathcal{S}^2 - \overline{X})^9 - n^9\Sigma_i X_i^9\Sigma_i X_i^9 - 2 n \Sigma_i X_i^9 \Sigma_i X_i (\Sigma_i X_i \mid n-1) \mid [\Sigma_i X_i (\Sigma_i X_i + n - 1)]^2$ as the sum of four terms I + II + III + IV, with

$$I = (n^2 - 2n)\ (\Sigma_i X_i^4 + \Sigma_{i \neq j} X_i^2 X_j^2)$$

$$II = -\ 2n\ \Sigma_i X_i^2\ \Sigma_{i \neq j} X_i X_j$$

$$III = -2n(n-1)\ (\Sigma_i X_i^3 + \Sigma_{i \neq j} X_i^2 X_j)$$

$$IV = Y^2(Y + n - 1)^2$$

where $Y = \Sigma_i X_i$ is Poisson with parameter $n\lambda$ (Haight 1967, p. 71). The expected values of I, III, and IV are simple to compute:

$$E(I) = (n^2 - 2n)n\ (\lambda + 7\lambda^2 + 6\lambda^3 + \lambda^4)$$

$$+ (n^2 - 2n)n(n-1)\ (\lambda^2 + 2\lambda^3 + \lambda^4)$$

$$E(III) = -2n^2(n-1)\ (\lambda + 3\lambda^2 + \lambda^3) - 2n^2(n-1)^2(\lambda^2 + \lambda^3)$$

$$E(IV) = EY^4 + 2(n-1)EY^3 + (n-1)^2EY^2$$

$$= n^4\lambda^4 + (2n^4 + 4n^3)\lambda^3 + (n^4 + 4n^3 + 2n^2)\lambda^2 + n^3\lambda.$$

To compute the expected value of term II we note firstly that there are $2n(n-1)$ terms having the form $X_i^3 X_j$ with $i \neq j$, and that there are $n(n-1)(n-2)$ terms having the form $X_i^2 X_j X_k$ with $i \neq j$, $j \neq k$, and $k \neq i$. Therefore

$$E(\mathrm{II}) = -4n^2(n-1)(\lambda^2 + 3\lambda^3 + \lambda^4) - 2n^2(n-1)(n-2)(\lambda^3 + \lambda^4).$$

Now, we can write the expected value of $n^2(n-1)^2(S^2 - \overline{X})^2$ as $E(\mathrm{I} + \mathrm{II} + \mathrm{III} + \mathrm{IV}) = \alpha_4 \lambda^4 + \alpha_3 \lambda^3 + \alpha_2 \lambda^2 + \alpha_1 \lambda$ with

$$
\begin{aligned}
\alpha_4 &= (n^2 - 2n)n + (n^2 - 2n)n(n-1) - 4n^2(n-1) - 2n^2(n-1)(n-2) + n^4 \\
&= n^3 - 2n^2 + n^4 - 2n^3 - n^3 + 2n^2 - 4n^3 + 4n^2 - 2n^4 + 2n^3 \\
&\qquad + 4n^3 - 4n^2 + n^4 = 0 \\
\alpha_3 &= 6(n^2 - 2n)n + 2(n^2 - 2n)n(n-1) - 2n^2(n-1) - 2n^2(n-1)^2 \\
&\qquad + 2n^4 + 4n^3 - 12n^2(n-1) - 2n^2(n-1)(n-2) = 0 \\
\alpha_2 &= 7(n^2 - 2n)n + (n^2 - 2n)n(n-1) - 6n^2(n-1) - 2n^2(n-1)^2 \\
&\qquad + n^4 + 4n^3 + 2n^2 - 4n^2(n-1) = 2n^2(n-1) \\
\alpha_1 &= (n^2 - 2n)n - 2n^2(n-1) + n^3 = 0,
\end{aligned}
$$

and thus, we have shown $E(\mathrm{I} + \mathrm{II} + \mathrm{III} + \mathrm{IV}) = \alpha_4 \lambda^4 + \alpha_3 \lambda^3 + \alpha_2 \lambda^2 + \alpha_1 \lambda = 2n^2(n-1)\lambda^2$, which completes the proof.

By replacing the variance of $(S^2 - \overline{X})$ with its corrected version we achieve the surprisingly simple test statistic (λ estimated by $\overline{X}$)

$$O_T^{(\mathrm{new})} = \sqrt{(n-1)/2}\,(S^2 - \overline{X})/\overline{X} = \sqrt{(n-1)/2}\,(S^2/\overline{X} - 1)$$

which is equivalent to the more popular *index of dispersion* defined as $(n-1)\,S^2/\overline{X}$ for which Hoel (1943) argues that the χ^2 with $(n-1)$ degrees of freedom gives a good approximation. To see the equivalence more completely we look at the corresponding standardization of the index of dispersion and we achieve the remarkable fact

$$[(n-1)S^2/\overline{X} - (n-1)]/\sqrt{2(n-1)} = O_T^{\mathrm{new}}.$$

Example 4.1: We conclude the discussion on this special overdispersion test with an example leading to completely different outcomes for O_T and O_T^{new}. The data presented in Table 4.1 are the number of daily death cases of women for the year 1989 in Berlin (western districts) with brain vessel disease (International Classification of Diseases 430–438) as cause of death.

death cases per day	0	1	2	3	4	5	6	7	8	9	10	11	12	13	14
frequency		1	4	15	31	39	55	54	49	47	31	16	9	8	4	3

Table 4.1. Cases of female death from brain vessel disease in Berlin, 1989

The statistics are $n = 366$, $\bar{x} = 6.3634$, $s^2 = 6.8238$, indicating a slight overdispersion of $s^2 - \bar{x} = 0.4604$, which leads to values of $O_T = 2.2706$ and $O_T^{new} = 0.9787$, leading to different test decisions.

As has been discussed already on Example 2.4, frequently a specific form of Poisson kernel is of interest. Observed death counts x are related to expected death counts e, where e is computed as $n_1\mu_1 + n_2\mu_2 + \ldots + n_j\mu_j$, where n_j is the size of age-stratum j in the study population and μ_j is the mortality rate in age-stratum j in a *reference population*. Consequently, we are led to a specific form of the Poisson kernel: $f(x, \lambda, e) = \exp(-\lambda e)\,(\lambda e)^x/x!$, where λ is the expected value of the standardized mortality ratio SMR, defined as SMR $= x/e$.

Quite similar to (4.6) we find here that

$$\mathrm{Var}\left\{ \frac{(X - \lambda e)^2}{e} - X/e \right\} = 2(\lambda e)^2/e^2 = 2\lambda^2 \tag{4.7}$$

and consequently,

$$\mathrm{Var}\left\{ \frac{1}{n}\sum_{i=1}^{n}\left[\frac{(X_i - \lambda e_i)^2}{e_i} - X_i/e_i \right] \right\} = 2\lambda^2/n .$$

The appropriately standardized statistic is

$$\sqrt{n/2}\,\frac{1}{n}\sum_{i=1}^{n}\left[\frac{(X_i - \lambda e_i)^2}{e_i} - X_i/e_i \right]/\lambda$$

and if λ is replaced by its arithmetic mean estimate $\bar{\lambda} = (1/n)\sum_{i=1}^{n} X_i/e_i$:

$$O_T^{new} = \sqrt{(n-1)/2}\,[S^2/\bar{\lambda} - 1] \tag{4.8}$$

with $S^2 = 1/(n-1)\sum_{i=1}^{n}(X_i - \bar{\lambda}e_i)^2/e_i$ which leads to the well-known Potthoff-Whittinghill statistic $\sum_{i=1}^{n}(X_i - \bar{\lambda}e_i)^2/\bar{\lambda}e_i$ (Potthoff and

Wittinghill 1966). Evidently, O_T^{new} is equivalent to $\sum_{i=1}^{n}(X_i - \bar{\lambda}e_i)^2/\bar{\lambda}e_i$ which is approximately χ^2 with $(n-1)$ degrees of freedom under homogeneity.

Example 4.2: We look at the data set of SMR-values of Martuzzi and Hills (1995) again, which consisted of small area data on perinatal mortality in a health region in England consisting of 515 areas. We find that $O_T^{new} = 4.2451$ indicating a significant overdispersion as already mentioned in Example 2.4.

4.2 The problem of testing for the number of components

Let $f(x, P_k) = f(x, \lambda_1)p_1 + \ldots + f(x, \lambda_k)p_k$ denote the mixture density as before though the number of components is now explicitly indexed in $f(x, P_k)$. Accordingly the log-likelihood is denoted as $l(P_k) = \Sigma_i \log f(x_i, P_k)$. We are interested in testing the hypothesis

$$H_0: \text{number of components} = k$$

against

$$H_1: \text{number of components} = k + 1$$

and use the *likelihood-ratio test* by means of the statistic

$$\text{LRS (Likelihood Ratio Statistic)} = 2 \log \xi_n = 2 \times [l(\hat{P}_{k+1}) - l(\hat{P}_k)].$$

Here $\hat{P}_{k+1}$ is the nonparametric maximum likelihood estimator under H_1 and $\hat{P}_k$ is NPMLE under H_0. The log-likelihood ratio test statistic $2 \log \xi_n$ generally has an asymptotic χ^2-distribution with d degrees of freedom, where the degrees of freedom, d, equal the difference between the number of parameters under the alternative and null hypothesis (Cox and Hinkley 1974, p. 323). In the case of univariate λ, this would imply a χ^2-distribution with 2 df for the test of H_0 and H_1 above. However, this theory is known to fail for the mixture problem (Titterington *et al.* 1985, p. 154). The explanation is that the null hypothesis does not lie in the interior of the parameter space. Recently, Goffinet *et al.* (1992) found the exact limiting distribution in several problems involving the normal distribution, but under the assumption that the weight parameters are known under the alternative hypothesis. In general, however, there is little known other than that the standard theory does not apply.

In statistical applications of mixture models the problem is still often ignored. One example is given by Gibbons *et al.* (1990), who use a mixture of Poissons in modeling suicide surveillance. They argue that a χ^2 distribution with 1 df for $2 \log \xi_n$ can be used for testing a one-component against a two-component model if n is beyond 20 or 30, although they point out in the same paper that the boundary condition is violated. The problem is not restricted to discrete mixtures only. It applies as well in the context of parametric mixtures. Martuzzi and Hills (1995) use the Poisson-Gamma mixture approach to model heterogeneity in small-area health data. They compare the model with the homogeneous Poisson model (heterogeneity variance $\tau^2 = 0$) and use the likelihood ratio test for this purpose in the conventional way. The hypothesis testing problem is easily formulated as H_0: $\tau^2 = 0$ vs H_1: $\tau^2 > 0$, and it is easily seen that the null-hypothesis is part of the boundary of the alternative hypothesis. Therefore, violations of standard conditions occur as well, though they are more easily dealt with than in the more general setting of finite mixture models.

A systematic investigation *by simulation* of the distribution of $2 \log \xi_n$ for various densities was not done until very recently. Thode *et al.* (1988) studied the null-distribution of the likelihood ratio statistic for the case that the alternative is a mixture of 2 normal densities with an *additional* free and common variance parameter. They conclude that the distribution of the likelihood ratio statistic is asymptotically χ^2 with 2 df, although convergence is rather slow. In Mendell *et al.* (1991) the asymptotic distribution of $2 \log \xi_n$ is studied *under the alternative hypothesis*. It is conjectured that the asymptotic distribution could be noncentral χ^2, possibly with 2 df.

We will first look at some simple analytical solutions, before we investigate the problem by means of simulation studies for densities from the exponential family.

4.3   Some analytical solutions

Suppose that X_1, ..., X_n are a random sample in which each X_i is binomial Bi(2, λ), e.g., f(x, λ) = $\binom{2}{x}(1 - \lambda)^x \lambda^{(2-x)}$, λ being the probability of a failure. This simple example is considered, since by a geometric analysis one is able to characterize exactly the structure of the likelihood ratio test. If we record $Y_0 =$ number of zeros, $Y_1 =$ number of ones, Y_2 = number of twos, then $(Y_0, Y_1, Y_2)^T$ has a multinomial distribution with probabilities $(\alpha_0, \alpha_1, \alpha_2)^T = (\lambda^2, 2\lambda(1 - \lambda), (1 - \lambda)^2)^T$. This vector is in the probability simplex $\{(\alpha_0, \alpha_1, \alpha_2)^T \mid \alpha_i \geq 0, \alpha_0 + \alpha_1 + \alpha_2 = 1\}$. We can graphically reproduce this simplex in two dimensions by omitting the unessential last coordinate, giving us $\Sigma_2 = \{(\alpha_0, \alpha_1)^T \mid \alpha_0 \geq 0, \alpha_1 \geq 0, \alpha_0 + \alpha_1 = 1\}$ (see Figure 4.2). The set of binomial probabilities $\Gamma = \{(\lambda^2, 2\lambda(1 - \lambda))^T \mid \lambda \in [0,1]\}$ forms a curve which connects the vertices $(0, 0)^T$ and $(1, 0)^T$. *This curve represents the multinomial probabilities allowable under the null hypothesis of* $k = 1$. We can identify each value of λ with a point on this curve. The alternative hypothesis consists of densities having the form $p_1 f(x, \lambda_1) + p_2 f(x, \lambda_2)$, with $p_1 + p_2 = 1$. In our plot (Figure 4.2), such a density corresponds to a convex combination of the two points on Γ corresponding to λ_1 and λ_2, with weights p_1 and p_2. Thus, it is clear that in this case, the alternative hypothesis yields as multinomial probabilities the entire convex hull of Γ, the shaded portion of Figure 4.2. As Figure 4.2 nicely shows, *the null hypothesis is part of the boundary of the alternative.* In this problem, we can *analytically* describe 2 log ξ_n as follows. If we place no restrictions on $(\alpha_0, \alpha_1, \alpha_2)^T$, then the maximum likelihood estimator is $(\hat{\alpha}_0, \hat{\alpha}_1, \hat{\alpha}_2)^T = (Y_0, Y_1, Y_2)^T/n$, and so corresponds to a point $(\hat{\alpha}_0, \hat{\alpha}_1)^T$ in Σ_2. The maximum likelihood

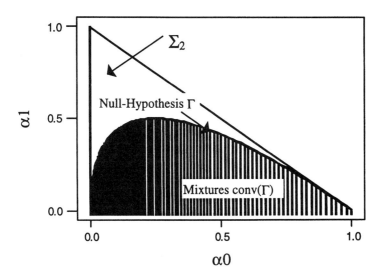

Figure 4.2. Null-hypothesis and alternative in the case of $f(x, \lambda) = \binom{2}{x}(1 - \lambda)^x \lambda^{2-x}$.

estimator under H_1: $k = 2$ must correspond to a point in the convex hull $conv(\Gamma)$ of Γ. First, if $(\hat{\alpha}_0, \hat{\alpha}_1)^T$ is in $conv(\Gamma)$, then this point must also be the maximum likelihood point under H_1. On the other hand, if $(\hat{\alpha}_0, \hat{\alpha}_1)^T \notin conv(\Gamma)$, then it can be shown that the maximum likelihood estimator under H_1 is on the boundary of $conv(\Gamma)$, hence in Γ, and so corresponds also to the maximum likelihood estimator under H_0, $\hat{\lambda} = (Y_1 + 2Y_2)/(2n)$. In this case, $2 \log \xi_n$ is zero.

What still remains to be answered is the question: what is the probability that $\xi_n = 1$? According to the above remarks, this is equivalent to $\hat{\alpha}$ lying above Γ or $\hat{\alpha}_1 \geq 2\sqrt{\hat{\alpha}_0}(1 - \sqrt{\hat{\alpha}_0})$ or $y_1 \geq 2\sqrt{y_0}(\sqrt{n} - \sqrt{y_0})$. Since $n\hat{\alpha}$ has the multinomial density

$$\binom{n}{y_0 y_1 y_2} \lambda^{2y_0} [2\lambda(1 - \lambda)]^{y_1} (1 - \lambda)^{2y_2},$$

this probability can be computed as

$$\eta_n(\lambda) = \Pr(\xi_n = 1) = \sum_{y_1 \geq 2\sqrt{y_0}(\sqrt{n} - \sqrt{y_0})} \binom{n}{y_0 y_1 y_2} \lambda^{2y_0} [2\lambda(1 - \lambda)]^{y_1} (1 - \lambda)^{2y_2}$$

Because of the asymptotic normality of $\hat{\alpha}$ we expect $\eta_n(\lambda)$ to converge to $\frac{1}{2}$. We can even say more about the form of convergence. Because of the convex curvature of Γ, it can be expected that convergence is from above. In principle, the distribution of $2 \log \xi_n$ conditional that $\xi_n > 1$ can be found in a similar way via the multinomial distribution. However, since for large sample size n the multinomial coefficients are expensive to compute we have simulated the conditional distribution of $2 \log \xi_n$. In Figure 4.3, an estimate of $\eta_n(\lambda)$ is shown for $n = 1000$, and a nonparametric estimate with associated 95% confidence interval of $\Phi_{\xi_n}^{-1}(0.95)$ and $\Phi_{\xi_n}^{-1}(0.99)$ is presented. $\Phi_{\xi_n}(x)$ is the conditional distribution function of $2 \log \xi_n$: $\Phi_{\xi_n}(x) = \Pr\{2 \log \xi_n \leq x \mid \xi_n > 1\}$. The confidence intervals were constructed in the usual manner using the normal approximation to the binomial. The estimate is based on a replication size of 10000. The solid lines in Figure 4.3 correspond to the 95th and 99th percentile of the χ^2-distribution with 1 df. For λ-values larger than 0.05, the χ^2-approximation appears to be rather satisfactory. Therefore, our analysis suggests that in this case the (unconditional) asymptotic distribution of $2 \log \xi_n$ is

$$\tfrac{1}{2} \chi^2_{(0)} + \tfrac{1}{2} \chi^2_{(1)},$$

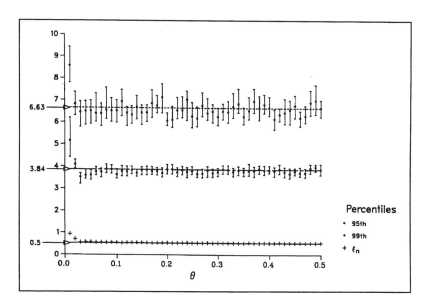

Figure 4.3. Binomial $(2, \lambda)$: η_r and percentiles of $2 \log \xi_r$ given that $\xi_r > 1$ for $n = 1000$.

where $\chi^2_{(0)}$ is the distribution with all its mass at zero. If we use the above geometric description and apply the asymptotic theory of Self and Liang (1987), then this example fits exactly in their Case 5.

Another example of this nature is the Poisson case with small λ-value. We consider $f(x, \lambda) = \exp(-\lambda)\lambda^x/x!$ with λ small, in the interval $[0, 0.1]$, say. Motivation for this assumption lies in the fact that then $P(X > 2) \approx 0$. Therefore, we can undertake an analysis similar to the one above, since the nonzero Poisson probabilities $[f(0, \lambda), f(1, \lambda), f(2, \lambda)]^T$ define a curve Γ in the two-dimensional simplex $\Sigma_2 = \{(\alpha_0, \alpha_1, \alpha_2)^T \mid \alpha_i \geq 0, \alpha_0 + \alpha_1 + \alpha_2 = 1 \}$. Again, we may consider only the first two coordinates: $\Gamma = \{(f(0, \lambda), f(1, \lambda))^T \mid \lambda \in [0, 0.1]\} \subseteq \Sigma_r$. Figure 4.4 demonstrates the geometry of the situation. Again, the null hypothesis is part of the boundary of the alternative. The complication is very similar to the binomial with sample size parameter 2, as discussed above.

What is different here is the way the event $\hat{\alpha}$ is above Γ is determined. We can write $\Gamma = \{\exp(-\lambda) (1, \lambda)^T \mid \lambda \in [0, 0.1]\}$. Thus the event $\hat{\alpha}$ is above Γ is equivalent to $\hat{\alpha}_1 \geq \hat{\alpha}_0 [-\log \hat{\alpha}_0]$ or, $y_1/y_0 \geq -\log(y_0/n)$ and can be computed via

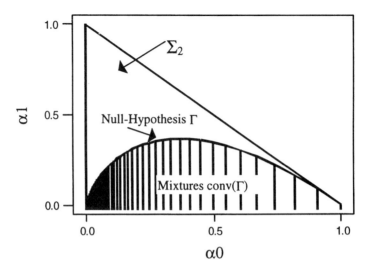

Figure 4.4. Null-hypothesis and alternative in the case of the Poisson with small λ.

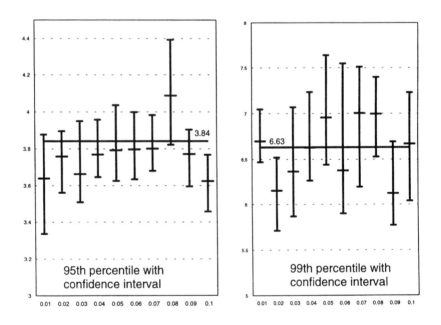

Figure 4.5. Poisson with small λ: percentiles of $2 \log \xi_n$ given that $\xi_n > 1$ for $n = 10,000$.

$$\eta_n(\lambda) = \Pr(\xi_n = 1) = \sum_{y_1/y_0 \geq -\log(y_0/n)} \binom{n}{y_0 y_1 y_2} f(0, \lambda)^{y_0} f(1, \lambda)^{y_1} f(2, \lambda)^{y_2},$$

the multinomial distribution. However, since the multinomial coefficients become expensive to compute for large n, we use simulation to find the conditional distribution of $2 \log \xi_n$ as well as $\eta_n(\lambda)$. In Figure 4.5, for $n = 10{,}000$, a nonparametric estimate with associated 95% confidence interval of $\Phi_{\xi_n}^{-1}(\alpha)$ for $\alpha = 0.95$ and $\alpha = 0.99$ is presented. Again, $\Phi_{\xi_n}(x)$ is the conditional distribution function of $2 \log \xi_n$: $\Phi_{\xi_n}(x)$ $= \Pr \{2 \log \xi_n \leq x \mid \xi_n > 1\}$. The estimate is based again on a replication size of 10 000. The solid lines in Figure 4.5 correspond to the 95th and 99th percentile of the χ^2-distribution with 1 df. For λ-values smaller than 0.10, the χ^2-approximation appears to be rather satisfactory. Therefore, our analysis suggests that in this case the (unconditional) asymptotic distribution of $2 \log \xi_n$ is

$$\tfrac{1}{2} \chi^2_{(0)} + \tfrac{1}{2} \chi^2_{(1)}$$

just as in the previous example.

Titterington, Smith, and Makov (1985, p. 152) provide a discussion on the issue and illustrate the problem in a very simple case, the mixture of two known densities:

$$f(x, p) = (1 - p) f_1(x) + p f_2(x) \tag{4.9}$$

where f_1 and f_2 are known densities. Let λ_1 and λ_2 be the expected values with respect to f_1 and f_2. Then

$$E(X) = (1 - p) \lambda_1 + p \lambda_2$$

leading to the *moment estimator*

$$\hat{p} = (\overline{X} - \lambda_1)/(\lambda_2 - \lambda_1) \tag{4.10}$$

Because of the restrictions on $p \in [0,1]$ only the truncated version $\hat{p}_+$ is meaningful where $\hat{p}_+ = \hat{p}$, if $\hat{p} \in (0, 1)$, $\hat{p}_+ = 0$, if $\hat{p} \leq 0$, $\hat{p}_+ = 1$, if $\hat{p} \geq 1$. Suppose that testing is H_0: $p = 0$ versus H_1: $p > 0$, and further, that H_0 is true. Then, because of the asymptotic normality of the mean one can expect the estimator (4.10) to be normal with mean 0. Therefore, we can expect *truncation to zero* in 50% of all cases. This is

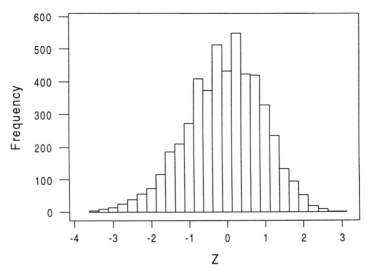

Figure 4.6. Distribution of $Z = \hat{p}/\text{s.e.}(\hat{p})$ for n = 1000 and mixture of two Poisson with $\lambda_1 = 1$ and $\lambda_2 = 4$

illustrated in Figure 4.6 for a mixture of two Poissons with mean 1 and mean 2, respectively. Note that the variance of $\hat{p}$ is given by $1/(\lambda_2 - \lambda_1)^2 \, [(1 - p)\sigma_1^2 + p\sigma_2^2 + (1 - p)(\lambda_1 - E(X))^2 + p \, (\lambda_2 - E(X))^2]/n$, where σ_i^2 is the variance w.r.t. f_i, $i = 1, 2$. If we use the estimator p_+ instead of the maximum likelihood estimator in the likelihood ratio test, it is quite clear that in about 50% of all cases the likelihood ratio statistic $2 \log \xi_n$ is zero, since the likelihoods under the null hypothesis and the alternative agree completely if $\hat{p} \leq 0$. The point is similar for the maximum likelihood estimator though it is more masked since it is one feature of the EM algorithm to keep the constraint $0 \leq \hat{p} \leq 1$. However, if different algorithms are used for determining the maximum likelihood estimator in this case such as Newton–Raphson or *regula falsi*, then "maximum likelihood estimators" in the sense of solutions of the score equation will occur on both sides of H_0: $p = 0$. Also, in this case $2 \log \xi_n$ is distributed as $\frac{1}{2} \chi^2_{(0)} + \frac{1}{2} \chi^2_{(1)}$.

Feng and McCulloch (1992) propose a different method for coping with the problem in this simple boundary situation. The idea is to extend the log-likelihood to the unbounded parameter space and, by this, retain the conventional properties of the likelihood ratio test. To demonstrate this idea in the simple situation above we consider $f(x, p) = (1 - p) \, f_1(x) + p \, f_2(x)$ and the restricted log-likelihood $l(p) = \Sigma_i \log f(x_i, p)$ with $p \in [0, 1]$. Now we extend $l(p)$ to the whole real line by $l^*(p) = \Sigma_i \log f^*(x_i, p) I_{f^*(x_i, p)}$ for all p, where I_z is the indicator function $I_z = 1$ if $z > 0$, and $I_z = 0$ otherwise. $f^*(x, p)$ is the extension of $f(x, p) = (1 - p) f_1(x) + p f_2(x)$ for any real number p. Then, under regularity conditions, Feng and McCulloch show that $2 \log \xi_n^* = 2$ $l^*(p) - 2 l^*(0)$ follows a χ^2 distribution with 1 degree of freedom, asymptotically. In fact, the example above is the first example presented in their Figure 1.

Evidently, the key idea of their approach is to allow values outside the restricted parameter space by extending the log-likelihood and therefore avoid the null hypothesis being part of the boundary of the alternative. This idea works in the context of mixtures as long as there is a *unique* way of embedding the null hypothesis into the alternative. However, consider $f(x) = (1 - p) \phi(x) + p \phi(x - \lambda)$, which will be discussed in detail further below. Obviously, the hypothesis $f(x) = \phi(x)$ can be embedded in two ways: $p = 0$ and λ arbitrary, or $\lambda = 0$ and p arbitrary.

The following example will enlighten the problem in further detail. We consider $f(x) = (1 - p) \phi(x) + p \phi(x - \lambda)$ and the hypothesis H_{01}: $f(x) = \phi(x)$ versus H_1: $p \neq 0$, $\lambda \neq 0$, where ϕ is the standard normal density. The problem is mentioned and discussed in Cheng and Traylor (1995), among other problems of non-standard inference. Following Hartigan (1985) the authors noted that the null distribution of $2 \log \xi_n$ is not only not χ^2 but actually converges to ∞. However, if we simulate the null distribution of $2 \log \xi_n$ in this case, for $n = 10,000$ we find the distribution given in Figure 4.7. There is almost complete agreement with the χ^2-distribution with 1 df, besides a small positive probability $\Pr\{2 \log \xi_n = 0\}$. The simulation is based on a replication size of 5000. The EM algorithm was used to find the maximum likelihood estimators, and as starting values the *moment estimators* were used that solve the moment equations $E(X) = p\lambda = \overline{X}$ and $\text{Var}(X) = (1 - p)p\lambda^2 + 1 = S^2$, where $\overline{X}$ and S^2 are the sample mean and sample variance, respectively. We doubt that the simulations are incorrect, since they have been confirmed independently; nor do we imply that the result of Hartigan (1985) is incorrect. However, our result certainly throws some light on the slowness of convergence.

There is a *second* form of null hypothesis H_{02} given by the boundary point $p = 1$ which often is of more interest in applications. It expresses the fact that an extra portion of 0s (non-cases, non-responders, etc.) might occur, and if the null-hypothesis is correct, this is not present and a simple mean model can be fitted to the data. In contrast with H_{01} one parameter, λ, must be estimated under H_{02}. Figure 4.8 represents the simulated null distribution of $2 \log \xi_n$ for $n = 10,000$.

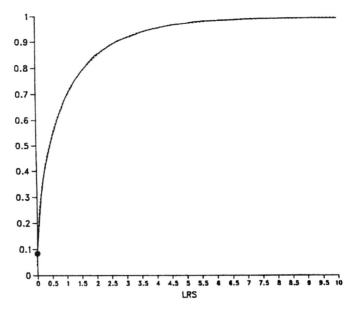

Figure 4.7. Simulated distribution function (solid line) of $2 \log \xi_n$ under H_{01} and χ^2-CDF (dashed line) with 1 degree of freedom for $n = 10{,}000$.

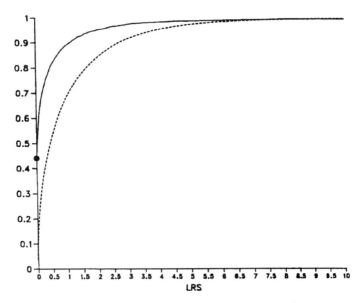

Figure 4.8. Simulated distribution function (solid line) of $2 \log \xi_n$ under H_{02} and χ^2-CDF (dashed line) with 1 degree of freedom for $n = 10{,}000$.

Clearly, the distribution of $2 \log \xi_n$ is shifted to the *left* (for any sample size) and not to the *right* as we might have expected. For further details see Böhning and Dietz (1995).

4.4   Simulation and bootstrap solutions

We consider now a general one-parametric density $f(x, \lambda)$. Unfortunately, the results of Section 4.3 do *not* generalize. A simple geometric characterization is no longer possible. However, with the tool of the *general mixture maximum likelihood theorem* (Theorem 2.1), it is possible to identify the cases for which $\xi_n = 1$ rather easily.

Recall that $\xi_n = 1$ is equivalent to $l(\hat{P}) = \sup_P l(P) = \sup_\lambda l(\lambda) = l(\hat{\lambda})$. Here, $\hat{P}$ and $\hat{\lambda}$ are the maximum likelihood estimators under the alternative and null hypotheses, respectively. The general mixture maximum likelihood theorem states that P^* is the unrestricted nonparametric maximum likelihood estimator if and only if $\sup_\lambda d(\lambda, P^*) = 1$. Because of (2.3) we have

$$n(\sup_\lambda d(\lambda, P) - 1) \geq l(P^*) - l(P) \geq 0$$

for any P from the probability simplex. Now in particular this inequality chain implies that

$$n(\sup_\lambda d(\lambda, \hat{\lambda}) - 1) \geq l(P^*) - l(\hat{\lambda}) \geq 0.$$

It follows that

$$n(\sup_\lambda d(\lambda, \hat{\lambda}) - 1) = 0 \text{ if and only if } l(P^*) - l(\hat{\lambda}) = 0.$$

Noting that $l(P^*) \geq l(\hat{P}) \geq l(\hat{\lambda})$, it is clear that $\sup_\lambda d(\lambda, \hat{\lambda}) = 1$ implies $l(\hat{P}) = l(\hat{\lambda})$, and so $\xi_n = 1$. On the other hand, if $l(\hat{P}) > l(\hat{\lambda})$, then $n(\sup_\lambda d(\lambda, \hat{\lambda}) - 1) > 0$ or $\sup_\lambda d(\lambda, \hat{\lambda}) > 1$. We note that with this procedure *no algorithm* is required[*] to determine whether $\xi_n = 1$.

For $\sup_\lambda d(\lambda, \hat{\lambda}) > 1$ the computation of $2 \log \xi_n$ is more difficult, and the use of algorithmic methodology cannot be avoided. Although

---

[*] Of course, it might be argued that the maximum of $d(\lambda, \hat{\lambda})$ needs to be determined. If $\hat{\lambda}$ does not coincide with the nonparametric maximum likelihood estimator the function $d(\lambda, \hat{\lambda})$ will have more than 1 local maximum and, therefore, a global search strategy is necessary.

many algorithms have been developed for computing the nonparametric maximum likelihood estimate of the mixing distribution (see Chapter 3), the EM algorithm is still the simplest technique to compute the maximum likelihood estimate when the support size is fixed (as it is here with $k = 2$). A key difficulty in mixture computations with fixed support size is the possible existence of multiple modes. In our simulations, we used starting values under the alternative $p_1 = p_2$ and $\lambda_1 = x_{(1)}$ and $\lambda_{(2)} = x_{(n)}$, since well separated values have often turned out to be a good strategy for avoiding local maxima which are not global ones – at least in univariate problems (Böhning et al. 1992). We note that computational error in the sense of not finding the global maximum would have the effect of biasing our percentiles downward. In any case, it is important to realize that not only the algorithm is important, but also the *choice of the initial values* might be relevant for determining the value of $2 \log \xi_n$.

In the following some results are put together for various component densities. Details can be found in Böhning et al. (1994). Table 4.2 presents some selected percentiles averaged over the 10 parameters under consideration.

sample size	95th percentile	99th percentile
100	4.01	7.15
1000	3.86	7.15
10,000	2.62	5.44
χ^2_2	5.99	9.21

Table 4.2. Selected percentiles of $2 \log \xi_n$ for Poisson distribution (averaged over parameter $\lambda = 1, 2, ..., 10$)

Next, we study — as a first example with a continuous sample space — the exponential distribution. Seidel et al. (1997, 1999) compute selected percentiles of $2 \log \xi_n$ which are given in Table 4.3. Here only the population with $\lambda = 1$ needs to be considered, since by a location invariance argument given in Seidel et al. (1997, 1999) $2 \log(\xi_n)$ does not depend on parameter λ under H_0.

sample size	95th percentile	99th percentile
100	4.79	8.02
1000	5.08	8.35
10,000	4.91	10.35
χ^2_2	5.99	9.21

Table 4.3. Selected percentiles of $2 \log \xi_n$ for exponential distribution ($\lambda = 1$)

We consider the normal density with known variance. Again, only the case $\lambda = 0$ needs to be considered under H_0. Selected percentiles can be found in Table 4.4.

sample size	95th percentile	99th percentile
100	3.50	6.63
1000	2.37	5.14
10,000	2.23	5.06
χ^2_2	5.99	9.21

Table 4.4. Selected percentiles of 2 log ξ_n for normal distribution ($\lambda = 0$) with known variance

Finally, for the normal with unknown variance we present selected percentiles based on the work of Thode *et al.* (1988).

sample size	95th percentile	99th percentile
100	6.68	10.33
1000	6.16	9.04
χ^2_2	5.99	9.21

Table 4.5. Selected percentiles of 2 log ξ_n for normal distribution ($\lambda = 0$) with unknown (and estimated) variance

In Tables 4.2–4.5 the corresponding percentiles of the χ^2-distribution with 2 degrees of freedom are given as well for comparison. As can be seen from the studies of the Poisson, the exponential, and the normal with *known* variance, the results are quite different from the conventional expectation. Only in the case where the variance parameter is estimated do the simulated results appear to converge to the conventionally expected ones.

Although the simulations given above throw some light on the distributional properties of the likelihood ratio test, they do not cover all potential situations which might occur. Therefore, McLachlan and Basford (1988) propose a computational technique which may validly be applied in *all* situations which may occur in practice. The approach is as follows. Let $f(x, P_k) = f(x, \lambda_1)p_1 + \ldots + f(x, \lambda_k)p_k$ denote the mixture density as before though the number of components is now explicitly indexed in $f(x, P_k)$. Accordingly, the log-likelihood is denoted as $l(P_k) = \Sigma_i$ log $f(x_i, P_k)$. We are interested in testing the hypothesis

$$H_0: \text{number of components} = k$$

against

$$H_1: \text{number of components} = k + 1.$$

Suppose we have found $\hat{P}_k$, the maximum likelihood estimate under the null hypothesis. Then a sample of size n is drawn from a population with mixture distribution $\hat{P}_k$ leading to a *bootstrap* sample x_1^*, ..., x_n^*. For this bootstrap sample the maximum likelihood estimate under the null hypothesis $\hat{P}_k$ as well as under the alternative $\hat{P}_{k+1}$ is determined and the likelihood ratio $2 \log \xi^*_n$ calculated. This is done B times and the B replicated values of $2 \log \xi^*_n{}^{(1)}$, ..., $2 \log \xi^*_n{}^{(B)}$ provide an estimate of the null distribution of the likelihood ratio from which the percentiles of interest can be calculated. A theoretical justification of the technique is given in Feng and McCulloch (1996).

We would like to point out that it is also possible to apply the nonparametric bootstrap approach to investigate the variability of the number of components k. The distribution of the number of components k may be obtained applying the mixture algorithm B times to bootstrap samples $(x_1^*, ..., x_n^*)$ obtained from the original sample with replacement.

As an example (to demonstrate the alternative way of illustrating the variability in the estimate $\hat{k}$), we provide the bootstrap distribution of $\hat{k}$ for a data set discussed in Richardson and Green (1997), the so-called acid data, obtained after $B = 5000$ replications. The details are given in Table 4.6, with f(k) denoting the relative frequency with which k components have been estimated. As can be seen in the table the mass of the distribution is lying on $k = 2$ and $k = 3$. Here the nonparametric mixture approach would indicate heterogeneity consisting of at least two components. Applying the likelihood ratio test of H_0: $k = 2$ against H_1: $k = 3$ leads to rejection of H_1 with a value of $2 \log \xi_n = 5.542$, which is nonsignificant on the 5% level according to simulation studies of Thode *et al.* (1988) transferring their results to our situation (see Table 4.5 and also Schlattmann and Böhning (1997) for further discussion).

f(1)	f(2)	f(3)	f(4)	f(5)	f(6)	f(7)	f(8)
0.0084	0.544	0.4356	0.0068	0.0034	0.0012	0.0002	0.0002

Table 4.6. Nonparametric bootstrap distribution of $\hat{k}$ for acid data based on 5000 replications

Example 4.3: In this final example we look at a simulation study to investigate the number of components estimator $\hat{k}$ inherently involved in the nonparametric maximum likelihood estimator of the mixing distribution. Let the component density be Poisson and $k = 2$ with $\lambda_1 = 1.25$ and $\lambda_2 = 2.66$ with weights $p_1 = 0.36$ and $p_2 = 0.64$. These values are in fact the maximum likelihood estimates for the *London Times Death Notices Data* presented in Table 1.3. Samples of sizes $n = 100, 1000, 2000,$ and 10 000 were drawn from this population and for each sample the nonparametric maximum likelihood estimate of P has been found including an estimate of the number of components k. This was done for each sample size 5000 times (replication size). The histogram of the 5000 $\hat{k}$-values is given in Figure 4.9 for each of the four sample sizes. Evidently, the shape of the distribution of $\hat{k}$ is similar for all sample sizes. As expected, the distribution of $\hat{k}$ is skewed to the right with some positive bias. On the other hand, the range of $\hat{k}$ is limited (maximum is always below 8) even for large sample sizes. Note that if one must decide upon the number of components, one would conventionally follow the mode (which value of k is receiving the largest weight), and in this example in all cases one would come up with the correct choice. Certainly, these issues have to be investigated in a refined manner in future investigations.

A variety of other techniques have been suggested for identifying the number of components of a mixing distribution including a variety of *graphical techniques* (Titterington, Smith, and Makov 1989; Roeder 1994; Fowlkes 1983; see also Lindsay and Roeder 1992, 1993; Böhning 1991).

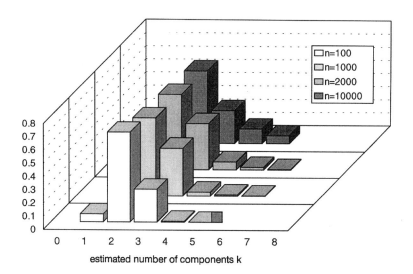

Figure 4.9. Histogram of the distribution of the estimated number of components.

CHAPTER 5

C.A.MAN-application: meta-analysis

5.1 The conventional approach

Meta-analysis can be defined as the quantitative analysis of a variety of single study results with the intention of an integrative presentation. See also Dickersin and Berlin (1992), Petitti (1994), Last (1995), Bailar (1995), or Jones (1995). Often in epidemiology or clinical trials we have as a measure of interest the *Odds-Ratio* Ψ, or equivalently the log (*Odds-Ratio*) $\lambda = \log(\Psi)$.* Then the following situation forms the basis for any meta-analysis. We have n independent studies (cohort, case-control) with estimates: $\hat{\lambda}_1$, ..., $\hat{\lambda}_n$ from which a *pooled estimate* $\hat{\lambda}_{\text{pool}} = w_1 \hat{\lambda}_1 + ... + w_n \hat{\lambda}_n$ is computed. The weights w_j are frequently chosen proportional to 1/var ($\hat{\lambda}_j$). There exists an extensive debate on the pros and cons of meta-analysis (see the review article of Dickersin and Berlin 1992).** On the critical side, it is usually mentioned that different studies might vary in their study quality. Some studies might have been done with greater care than others. For example, a case-control study might have been organized to sample study controls in a less biasing way than in a different case-control study. Also, frequently the problem of *publication bias* is mentioned. Recently, there have been a variety of papers devoted to this problem (Begg and Berlin 1988, Dear and Begg 1992, Hedges 1992). Publication bias refers to a certain type of selection bias, which can occur if studies with significant results are more likely to be published than other studies, leading to an *overestimation* of the effect size.*** In addition

---

* This is just one example of a measure of interest. Other measures are of interest, such as the standardized difference or the correlation coefficient.

**—Whenever an interesting new method is criticized for its misuses, guidelines are developed for how this new method should be used correctly. An example in this context is given by Cook *et al.* (1995).

*** In addition to modeling publication bias (and then correcting for it) one can try to establish study registrars such as is done with the *Cochrane Collaboration* (Chalmers 1993) for clinical trials.

to the many arguments in favor of meta-analysis, including an enormous improvement of sample size, most importantly it seems that it is becoming more and more part of the scientific method of providing evidence in favor or against a certain hypothesis or argument (see Figure 5.1). According to Petitti (1994) one can distinguish *four* steps in meta-analysis. In *step 1* all studies with relevant data are identified applying *systematic* and *explicit* search procedures for study identification. In this respect, it is clearly different from the qualitative literature review which is solely in the hands of the individual researcher. Systematic search procedures are used to avoid the occurrence of any selection bias and explicit search procedures are used to allow reproducibility. One should point out that this step usually has a process character. A starting point might be the collection of personal literature records, then the search is continued with the help of computerized literature databases such as MEDLINE. After the title and abstract of each candidate publication have been investigated and irrelevant publications have been eliminated, the remaining publications are read completely for the identification of relevant information. References in *each* found publication are investigated for further unknown studies (cross-checking). Finally, the database (list

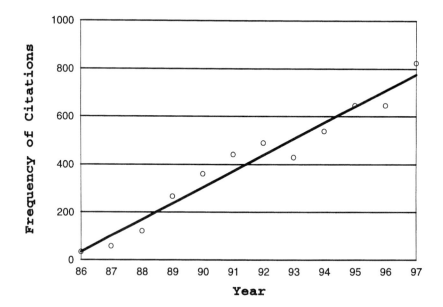

Figure 5.1. Demonstration of the increase of meta-analyses: MEDLINE-search with search term "Meta-Analysis."

of relevant publications) might be reviewed by an expert in the field. In *step 2* the inclusion and exclusion criteria are defined: not all studies can be allowed for the meta-analysis; for example, no non-experimental studies can be allowed in a meta-analysis of randomized trials. Similar to other conventional study designs the target here is to avoid bias at study selection and enhance reproducibility. *Examples* of inclusion or exclusion criteria include the study design, the time at study execution or publication, language of publication, multiple publications, the sample size, the comparability of treatment/exposure, or the completeness of information. In *step 3* the data on feasibility of study for the meta-analysis are identified (this is the information about meeting the inclusion and exclusion criteria, for example). After this has been done the creation and construction of the data base using the relevant statistical measures are started. Variables forming the data base could be *study identifier, effect measure of interest, study year, sample size, study type,* and similar variables of interest for the meta-analysis. Finally, in *step 4* the statistical analysis of the database is performed.

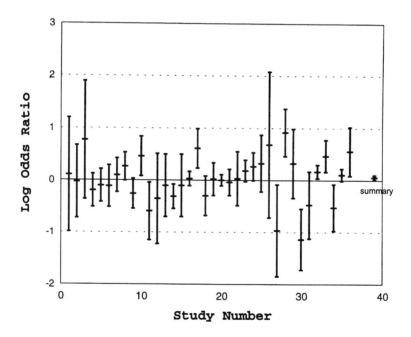

Figure 5.2. Effect estimates (log-odds ratios) of 36 studies on the relationship of hormone replacement therapy and breast cancer (after Sillero-Arenas *et al.* 1992).

Example 5.1: As one example* consider the meta-analysis provided by
Sillero-Arenas *et al.* (1992) on the relationship of hormone replacement
therapy and the occurrence of breast cancer. Studies were located by
MEDLINE, supplemented by a hand search of all references in the
articles located. Thirty-seven original studies were found: 23 case-control, 13 cohort, and one clinical trial (the latter has not been included
in the meta-analysis since it is not an observational study as the other
36). Figure 5.2 presents the effect estimates with pointwise 95% confidence intervals for 36 studies (case control and cohort).

5.2   Strategies of optimal pooling

The question arises of what to do with the individual effect measures
now available in the meta-analysis. The researcher is usually not
interested in the individual results, but in a more general message
bringing all individual results together. In the early days of meta-
analysis this was accomplished by simply pooling the n study estimates
$\hat{\lambda}_1, \ldots, \hat{\lambda}_n$ to form a *pooled estimate*

$$\hat{\lambda}_{\text{pool}} = w_1\hat{\lambda}_1 + \ldots + w_n\hat{\lambda}_n \tag{5.1}$$

where the weights w_i are non-negative and sum to 1: $w_i \geq 0$ for $i = 1,$
$\ldots, n$ and $w_1 + \ldots + w_n = 1$. Many choices of $w_1, w_2, \ldots, w_n$ are possible
and the question arises of which is the best one. This is answered by
the following theorem. Recall that $\text{var}(\hat{\lambda}_i) = \sigma_i^2$ for $i = 1, \ldots, n$. Also,
we assume for the remainder of this chapter that $\hat{\lambda}_1, \ldots, \hat{\lambda}_n$ are
independent realizations of the estimated measure of interest $\hat{\lambda}$.

Theorem 5.1: The variance of any pooled estimator (5.1) is given by
$w_1^2\sigma_1^2 + \ldots + w_n^2\sigma_n^2$. Then,

a) the *best pooled estimator* of form (5.1) (in the sense of minimal
variance) is given as

$$\hat{w}_i = \frac{1/\sigma_i^2}{\Sigma_j 1/\sigma_j^2} \quad \text{for } i = 1, \ldots, n \tag{5.2}$$

Study number	log-odds ratio	Variance
1	0.10436	0.299111
2	−0.03046	0.121392
3	0.76547	0.319547
4	−0.19845	0.025400
5	−0.10536	0.025041
6	−0.11653	0.040469
7	0.09531	0.026399
8	0.26236	0.017918
9	−0.26136	0.020901
10	0.45742	0.035877
11	−0.59784	0.076356
12	−0.35667	0.186879
13	−0.10536	0.089935
14	−0.31471	0.013772
15	−0.10536	0.089935
16	0.02956	0.004738
17	0.60977	0.035781
18	−0.30111	0.036069
19	0.01980	0.024611
20	0.00000	0.002890
21	−0.04082	0.015863
22	0.02956	0.067069
23	0.18232	0.010677
24	0.26236	0.017918
25	0.32208	0.073896
26	0.67803	0.489415
27	−0.96758	0.194768
28	0.91629	0.051846
29	0.32208	0.110179
30	−1.13943	0.086173
31	−0.47804	0.103522
32	0.16551	0.004152
33	0.46373	0.023150
34	−0.52763	0.050384
35	0.10436	0.003407
36	0.55389	0.054740

Table 5.1. Extracted data from Sillero-Arenas *et al.* (1992) on the relationship of hormone replacement therapy and breast cancer

b) Let $\hat{\lambda}_{best} = \Sigma_i \hat{w}_i \hat{\lambda}_i$, with weights given in (5.2). It is

$$\text{var}(\hat{\lambda}_{best}) = 1/\Sigma_j 1/\sigma_j^2 . \tag{5.3}$$

Proof. Because of the independence assumption the variance of any pooled estimator of form (5.1) is $\text{var}(w_1\hat{\lambda}_1 + ... + w_n\hat{\lambda}_n) = w_1^2\sigma_1^2 + ... + w_n^2\sigma_n^2 = f(w)$. To prove (a) this variance needs to be minimized with respect to $w_1, ..., w_n$ under the constraints $w_i \geq 0$ for $i = 1, ..., n$ and $\Sigma_i w_i = 1$. Evidently, f is a convex function. According to Corollary 2.1 in Chapter 2, $\hat{w} > 0$ is minimizing f if and only if $\partial f/\partial w_i(\hat{w}) = \Sigma_j \partial f/\partial w_j(\hat{w})\hat{w}_j$ for $i = 1, ..., n$. We have $\partial f/\partial w_i(w) = 2w_i\sigma_i^2$ and for $\hat{w}_i = 1/\sigma_i^2/\Sigma_j 1/\sigma_j^2$ in particular, $\partial f/\partial w_i(\hat{w}) = 2/\Sigma_j 1/\sigma_j^2$. This coincides with $\Sigma_j \partial f/\partial w_j(\hat{w})\hat{w}_j = 2/\Sigma_j 1/\sigma_j^2$ and ends the proof.

Example 5.2: We continue our discussion of Example 5.1. We find $\hat{\lambda}_{best}$ as 0.05598 with 95% C.I. [0.0118, 0.1002]. Retransformation to the relative risk scale provides $\hat{\lambda}_{best} = \exp(\hat{\lambda}_{best}) = 1.05758$ with 95% C.I. [1.01187, 1.10535], showing a small but significantly elevated relative risk.

This kind of analysis requires the availability of the effect measure *and* its variance. Often the variance is not directly available from the publication, but a *95% confidence interval* is provided. Assuming that it has been computed by means of $\lambda_{U, L} = \hat{\lambda}\pm 1.96\sqrt{\text{var}(\hat{\lambda})}$, where $\lambda_{U, L}$ are the upper and lower confidence limits, respectively, it is not difficult to see that the variance is provided by $\text{var}(\hat{\lambda}) = [(\lambda_U - \lambda_L)/(2 \times 1.96)]^2$. Often, instead of the confidence interval, a *P-value* is provided in the publication as $\Pr(Z \geq z)$, where $Z = \hat{\lambda}/\sigma(\hat{\lambda})$ with $\sigma(\hat{\lambda})$ being the standard error of estimate $\hat{\lambda}$ (e.g., $\text{var}(\hat{\lambda}) = \sigma^2(\hat{\lambda})$). Assuming a valid normal approximation the P-value can be computed as $1 - \Phi(z)$, or $\Phi(z) = 1 - P\text{-value}$, or $z = \Phi^{-1}(1 - P\text{-value})$. Here Φ is the cumulative distribution function of the standard normal. Since z is the observed value of $\hat{\lambda}/\sigma(\hat{\lambda})$, we can recompute var $(\hat{\lambda})$ as $\sigma^2(\hat{\lambda}) = [\hat{\lambda}/\Phi^{-1}(1 - P\text{-value})]^2$.

As mentioned above, *other* effect measures are commonly used in meta-analysis. In psychology the standardized difference is considered frequently. Let $\delta_i = (\mu_i^E - \mu_i^C)/\sigma_i$ be the population standardized difference of the *i*th study, where μ_i^E, μ_i^C, σ_i are the population mean in treatment, control group, and common standard deviation of the *i*th study. δ_i can be estimated as $\hat{\delta}_i = g_i = (\bar{x}_i^E - \bar{x}_i^C)/s_i$, where $\bar{x}_i^E$ and $\bar{x}_i^C$ are the sample means in the treatment and control group, respectively, with sample sizes n_i^E for treatment and n_i^C for control group. According to Hedges and Olkin (1985) two estimates of σ_i are commonly considered:

a) $s_i^2 = s_i^{C2} = \dfrac{1}{n_i^C - 1}\sum_{j=1}^{n_i^C}(x_{ij}^C - \bar{x}_i^C)^2$

b) $s_i^2 = \left[\sum_{j=1}^{n_i^C}(x_{ij}^C - \bar{x}_i^C)^2 + \sum_{j=1}^{n_i^E}(x_{ij}^E - \bar{x}_i^E)^2\right]/(n_i^C + n_i^E - 2)$

The estimate (a) is solely based on the control group (leading to the so-called *Glass's* Δ), whereas the estimate (b) is a pooled estimate (leading to the so-called *Cohen's d*). For further discussion see Rosenthal (1994). Unfortunately, $\hat{\delta}_i$ is a biased estimate of δ_i: $E(\hat{\delta}_i) \approx \delta_i + 3\delta_i/(4(n_i^E + n_i^C) - 9)$. On this basis a bias-corrected estimate $d_i = (1 - (3/(4(n_i^E + n_i^C) - 9))) (\bar{x}_i^E - \bar{x}_i^C)/s_i$ is considered. The exact value of $E(\hat{\delta}_i) = \delta_i / J(N_i - 2)$, with $N_i = (n_i^E + n_i^C)$, leading to the exact bias correction: $d_i = J(N_i - 2) (\bar{x}_i^E - \bar{x}_i^C)/s_i$. The biasing constant $J(m)$ cannot be given in closed form, but Hedges and Olkin provide a numerical table of it (see Table 5.2). It is clear that for larger sample sizes (above 10), the differences between $\hat{\delta}_i$ and d_i are negligible.

m	2	3	4	5	10
$J(m)$	0.5642	0.7236	0.7979	0.8408	0.9228

Table 5.2. Numerical values of biasing constant $J(m)$

Hedges and Olkin (1985) also provide an approximate expression for the variance of d_i:

$$\sigma^2(d_i) \approx \frac{n_i^E + n_i^C}{n_i^E n_i^C} + \frac{\delta_i^2}{2(n_i^E + n_i^C - 2)} \qquad (5.4)^*$$

which can be estimated by

$$\hat{\sigma}^2(d_i) \approx \frac{n_i^E + n_i^C}{n_i^E n_i^C} + \frac{d_i^2}{2(n_i^E + n_i^C - 2)} \qquad (5.5)$$

Note that if δ_i is close to zero, (5.4) is just the sum of the inverse sample sizes of treatment and control group.

---

* If s_i is computed solely on the basis of the control group the variance is $\sigma^2(d_i) \approx (n_i^E + n_i^C)/(n_i^E n_i^C) + \delta_i^2/(2n_i^C - 2)$ (Rosenthal 1994).

Example 5.3: Hedges and Olkin (1985) discuss a meta-analysis on the question of how education (open versus traditional) influences a student's self-concept. See Figure 5.3. Doing the pooled analysis we find d_{best} = $\Sigma_i \, \hat{w}_{i} d_i = 0.011$, $var(d_{best}) = 1/\Sigma_j \, (1/\sigma_j^2) = 0.0018$ and the 95% C.I.: $d_{best} \pm (2 \times \sqrt{1/\Sigma_j(1/\sigma^2)}) = (-0.072, 0.094)$, which can be also seen in Figure 5.3 on the right-hand side. Evidently, the summary estimator does not show any evidence of effect.

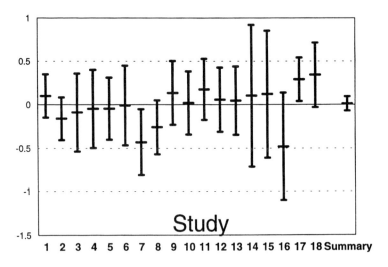

Figure 5.3. Standardized effect difference in 18 studies of the effect of open versus traditional education on student self-concept (after Hedges and Olkin, p. 111).

Another effect measure frequently used in the social sciences is the correlation coefficient $\rho_i = cov(X_i, Y_i)/ \sqrt{Var(X_i)Var(Y_i)}$ of two measurements, X_i and Y_i, in the i-th study. The natural estimate is $r_i = \hat{\rho}_i$, the sample correlation coefficient in the ith study. However, the normal approximation of the distribution of r_i is not good and the *Fisher-Transformation* $z_i = \frac{1}{2} \log((1 + r_i)/(1 - r_i))$ is used instead. The approximate variance of z_i is $var(z_i) \approx 1/(n_i - 3)$. Obviously, the optimally pooled estimate is given by weighting each study estimate z_i by its sample size.

Example 5.4: We consider a data set of seven studies also discussed in Hedges and Olkin (1985) on the relationship of teacher indirectness and student achievement. Here the effect measure is the correlation coefficient. We find $z_{best} = \Sigma_i (n_i - 3) z_i / \Sigma_i (n_i - 3) = 0.2896$, leading to $r_{best} = [\exp(2 \, z_{best}) - 1]/[\exp(2 \, z_{best}) + 1] = 0.2818$. Based on the variance

study i	N_i	r_i
1	15	−0.073
2	16	0.308
3	15	0.481
4	16	0.428
5	15	0.180
6	17	0.290
7	15	0.400

Table 5.3. Seven studies on the relationship of teacher indirectness and student achievement (after Hedges and Olkin 1985)

var(z_{best}) = 0.01136 we find the 95% C.I. as [0.0807, 0.4985], and on the original scale the 95% C.I. for ρ is [0.0805, 0.4610].

5.3   A likelihood approach

In this section we will try to give a justification of the pooled estimator by the maximum likelihood method. We recall the following facts for a random variable X with density f(x, λ). Let a random sample x_1, x_2, ..., x_n be given. Then the likelihood is the product f(x_1, λ) f(x_2, λ) × ... × f(x_n, λ) and, as a function of λ, it is called the *likelihood function* and denoted by L(λ). The maximum likelihood estimator $\hat{\lambda}_{\text{ML}}$ is defined as that value of λ with the following property: L($\hat{\lambda}_{\text{ML}}$) ≥ L(λ) for all λ. The maximum likelihood estimator is famous for its properties including consistency, asymptotic normality, and that its asymptotic variance is given through the Rao–Cramer-bound: −1/(d²/dλ²logL) (see Cox and Hinkley 1974 for details and further discussion). We apply this theory to our situation. The measure of interest λ is estimated in n studies leading to n estimates each being distributed $\hat{\lambda}_i \sim N(\lambda, \sigma_i^2)$, e.g., with common, unknown effect parameter and known variances σ_i^2, $i = 1, ..., n$. More precisely, for each estimate $\hat{\lambda}_i$ there is a density of normal type: $1/\sqrt{2\pi\sigma_i^2}\exp\{-\frac{1}{2}(\hat{\lambda}_i - \lambda)^2/\sigma_i^2\}$, leading to the following likelihood:

$$L(\lambda) = \prod_{i=1}^{n} \frac{1}{\sqrt{2\pi\sigma_i^2}}\exp\left\{-\frac{1}{2}(\hat{\lambda}_i - \lambda)^2/\sigma_i^2\right\}$$

or alternatively, to the following log-likelihood:

$$l(\lambda) \;=\; \log L(\lambda) \;=\; \sum_{i=1}^{n} -\log(\sqrt{2\pi\sigma_i^2}) - \frac{1}{2}(\hat{\lambda}_i - \lambda)^2/\sigma_i^2 \qquad (5.6)$$

The *score* is given as $l'(\lambda) = (d/d\lambda)l(\lambda) = \sum_{i=1}^{n}(\hat{\lambda}_i - \lambda)/\sigma_i^2$ leading to the score equation:

$$l'(\lambda) \;=\; 0 \quad \text{or} \quad \sum_{i=1}^{n} \hat{\lambda}_i/\sigma_i^2 \;=\; \lambda \sum_{i=1}^{n} 1/\sigma_i^2$$

and the maximum likelihood solution

$$\hat{\lambda}_{\mathrm{ML}} \;=\; \frac{\displaystyle\sum_{i=1}^{n} \hat{\lambda}_i/\sigma_i^2}{\displaystyle\sum_{i=1}^{n} 1/\sigma_i^2} \;=\; \hat{\lambda}_{\mathrm{best}}\,.$$

We note that the maximum likelihood estimate is just the optimally pooled estimate (5.2) from the previous section. Computation of the Rao–Cramer bound $-1/(d^2/(d\lambda)^2 \log L)$ leads to $l'(\lambda) = \sum_{i=1}^{n}(\hat{\lambda}_i - \lambda)/\sigma_i^2$, and further to $l''(\lambda) = -\sum_{i=1}^{n} 1/\sigma_i^2 = -1/\mathrm{Var}\,(\hat{\lambda}_{\mathrm{ML}})$.

Example 5.5: The likelihood principle can also be of use in situations where it is less clear how to proceed to achieve an overall estimate. Eaton (1995) discusses a meta-analysis of agoraphobia based on seven studies as given in Table 5.4. Agoraphobia* can be defined as space anxiety, as fear of being on public streets and squares or even to leave a protected room or house (apartment), the fear of being in lonely places; it can be described as an urgent attack with a strong desire to avoid the situation. Agoraphobia is considered to be an extreme form of phobia with respect to its disabling nature; the cause (for this strength) can be seen in the *diffuseness* of the situation that initiates the fear. It stands in contrast to other fears (such as fear of snakes, elevators, tunnels, high buildings, planes, etc.) which are limited and as such can be avoided.

Let $\hat{\lambda}_i$ denote the estimated prevalence in the ith study (number of cases y_i divided by number of study participants N_i), e.g., $\hat{\lambda}_i = y_i/N_i$. Then, under the rare disease assumption a suitable likelihood for the ith study

---

* In the Greek language *agora* means market square.

Study no.	Prevalence (per 10^3)	Sample size
1	56	14 436
2	57	1 366
3	69	1 551
4	29	3 258
5	21	3 134
6	36	1 966
7	53	8 098

Table 5.4. Seven studies on estimates of agoraphobia

is given through the Poisson distribution: $\exp\{-\lambda N_i\}\{\lambda N_i^{y_i}\}/y_i!$ which leads to a log-likelihood of

$$-\lambda N_i + y_i \log(\lambda) \tag{5.7}$$

where the data-part not involving λ has been ignored. If the full likelihood is considered we yield

$$l(\lambda) = -\lambda \, \Sigma_i N_i + \Sigma_i y_i \log(\lambda)$$

and as score $dl/d\lambda = -\Sigma_i N_i + \Sigma_i y_i / \lambda$. The score equation $dl/d\lambda = 0$ leads to the maximum likelihood estimate $\hat{\lambda}_{\mathrm{ML}} = \Sigma_i y_i / \Sigma_i N_i$. Its variance is given as $\mathrm{var}(\hat{\lambda}_{\mathrm{ML}}) = \mathrm{var}(\Sigma_i y_i / \Sigma_i N_i) = 1/(\Sigma_i N_i)^2 \Sigma_i \mathrm{var}(y_i) = 1/(\Sigma_i N_i)^2 \, \Sigma_i \, \lambda N_i = \lambda/\Sigma_i N_i$. To be of practical use the unknown parameter λ needs to be replaced by its estimate, leading to the estimated variance

$$\hat{\mathrm{var}}(\hat{\lambda}_{\mathrm{ML}}) = \frac{\Sigma_i y_i}{(\Sigma_i N_i)^2} \tag{5.8}$$

For the data of Table 5.4 we find $\hat{\lambda}_{\mathrm{ML}} = \Sigma_i y_i/\Sigma_i N_i = 48.89$ (per 1000 population) and $\mathrm{S.E.}(\hat{\lambda}_{\mathrm{ML}}) = \sqrt{\hat{\mathrm{var}}(\hat{\lambda}_{\mathrm{ML}})} = 1.2026$, leading to a 95% C.I. as $\hat{\lambda}_{\mathrm{ML}} \pm 2 \times \mathrm{S.E.}(\hat{\lambda}_{\mathrm{ML}}) = [46.49, 51.3]$.

5.4   Heterogeneity

One question of debate in meta-analysis is whether individual study estimates of effect can validly be pooled into a common estimate of effect. This is conveniently put as the question of *homogeneity* or *heterogeneity* of study results. Homogeneity is conventionally investigated by *diagnostic tests* such as the χ^2-test of homogeneity. This test can be simply described as

$$\chi^2_{(n-1)} = \sum_{i=1}^{n} (\hat{\lambda}_i - \hat{\lambda}_{\text{best}})^2 / \sigma_i^2$$

where the terms involved in the formula above are defined in (5.2). For the data on breast cancer and hormone replacement therapy in Example 5.1 we find a $\chi^2_{(35)} = 116.076$, which has a very small *P*-value on the χ^2-scale with 35 df. The χ^2-test of homogeneity has as an advantage mainly its simplicity. It can be compiled using a simple pocket calculator. It has been criticized recently in a number of papers, essentially on the grounds of its lack of power (see Harwell 1997, Hardy and Thompson 1997, Sánchez-Meca and Marín-Martínez 1997, Cornwall and Ladd 1993). If there is evidence for heterogeneity, then the problem remains how to proceed. The mixture approach provides an *elegant solution* for this problem in that it models the heterogeneity distribution in a nonparametric way.

The underlying assumption in most forms of meta-analysis — expressed also graphically in Figure 5.2 by computing symmetric confidence intervals — is that of a normal distribution for the effect estimate $\hat{\lambda}_i \sim N(\lambda_i, \sigma_i^2)$, with $\sigma_i^2 = \text{var}(\hat{\lambda}_i)$. Note that it is important to allow for different variances since the sample size will differ from study to study. We have seen in the preceding section that in the simplest case of *homogeneity* ($\lambda_1 = \lambda_2 = \ldots = \lambda_n = \lambda$) the MLE of λ corresponds to the pooled estimator. If the population is heterogeneous we must assume the existence of *subpopulations* with parameter λ_j receiving weight p_j for the jth subpopulation. This means that $\hat{\lambda}_i$ has a normal density of type: $f(\hat{\lambda}_i, \lambda_j) := 1/\sqrt{2\pi\sigma_i^2} \exp\{-\tfrac{1}{2} (\hat{\lambda}_i - \lambda_j)^2/\sigma_i^2\}$; *conditionally it is known* that $\hat{\lambda}_i$ has been generated from subpopulation j. However, this information is usually *unknown*. Therefore let Z be an unobserved or *latent* variable describing the population membership of each study. Then, the joint density $f(\hat{\lambda}_i, z)$ is given as

$$f(\hat{\lambda}_i, z) = f(\hat{\lambda}_i | z) f(z) = f(\hat{\lambda}_i; \lambda_z) p_z$$

where $f(\hat{\lambda}_i | z) = f(\hat{\lambda}_i; \lambda_z)$ is the density *given that $\hat{\lambda}_i$ is coming from a subpopulation with parameter λ_z* and $f(z) = p_z$ is the probability that the ith study is coming from subpopulation z.

Consequently, the *unconditional* density of $\hat{\lambda}_i$ is given as the marginal or mixture density over z ($= 1, 2, \ldots, k$):

$$\sum_{j=1}^{k} f(\hat{\lambda}_i; \lambda_j) p_j = \frac{1}{\sigma_i} \sum_{j=1}^{k} \varphi\left(\frac{\hat{\lambda}_i - \lambda_j}{\sigma_i}\right) p_j . \tag{5.9}$$

Here, φ is the standard normal density $\varphi(x) = \exp\{-x^2/2\}/\sqrt{2\pi}$. Note also that the number of subpopulations k is *not* assumed to be known. Thus, to allow for heterogeneity we are led again to a mixture model. What remains is that the mixing distribution $P = \begin{pmatrix} \lambda_1 \cdots \lambda_k \\ p_1 \cdots p_k \end{pmatrix}$ needs to be estimated! This can be accomplished by C.A.MAN which computes the maximum likelihood estimator of P by maximizing the log-likelihood

$$l(P) = \sum_{i=1}^{n} \log \sum_{j=1}^{k} f(\hat{\lambda}_i, \lambda_j) p_j$$

C.A.MAN provides the maximum likelihood estimate for P, namely, $\hat{P} = \begin{pmatrix} \hat{\lambda}_1 \cdots \hat{\lambda}_k \\ \hat{p}_1 \cdots \hat{p}_k \end{pmatrix}$ from which detailed knowledge on the cluster-structure involved in the meta-analysis can be gained. Some details on how this is done can be found in the next section.

5.5   C.A.MAN solution for modeling heterogeneity

We now take a look at several examples of meta-analysis for highlighting the modeling of heterogeneity.

Example 5.6: *(Example 5.3 continued)* The $\chi^2_{(17)}$ - statistic delivers a value of 22.72 with associated P-value 0.1586 (17df), indicating a rather small amount of heterogeneity in these 18 studies. Table 5.5 gives details of the analysis with C.A.MAN.

Parameter	Weights	Log-likelihood
k = 2		
−0.13932	0.4239	1.89838
0.11554	0.5761	
k = 1		
0.0106	1.0000	0.93321

Table 5.5. Nonparametric maximum likelihood estimate for meta-analysis of educational style (traditional versus conventional) on student self-concept

The package finds the nonparametric maximum likelihood estimator to consist of *two* components ($k = 2$), one having a negative mean difference, the other a positive mean difference of similar size and weight. A comparison with the homogeneous ($k = 1$) case is useful to determine if the found heterogeneity is significant. The difference in the log-likelihood value is below 1, which even on the corrected χ^2-scale is not significant (see Chapter 4).

We can conclude from this analysis that there is no evidence for unobserved heterogeneity and that the pooled estimator will most likely provide an adequate summary of the individual studies.

Example 5.7: The second example is based on a meta-analysis of 12 case-control studies on the relationship between the usage of oral contraceptives and the occurrence of breast cancer and goes back to Malone *et al.* (1993). Details on the C.A.MAN-analysis are given in Table 5.6. The results provide some evidence of heterogeneity consisting of two subpopulations, one having an elevated risk of exp(0.944) = 2.57 receiving weight 0.26. In other words, 26% of all studies belong to a group with an elevated risk of 2.57. The NPMLE of the heterogeneity distribution can be seen in Figure 5.4

Parameter	Weights	Log-likelihood
$k = 3$		
0.1803	0.7426	−9.91313
0.9384	0.2270	
0.9837	0.0304	
$k = 2$		
0.1803	0.7427	−9.91317
0.9441	0.2573	
$k = 1$		
0.3226	1.0000	−11.06471

Table 5.6. Nonparametric maximum likelihood estimate for meta-analysis of the usage of oral contraceptive and the occurrence of breast cancer

Example 5.8: The final example continues the discussion on the meta-analysis provided by Sillero-Arenas *et al.* (1992) on the relationship of hormone replacement therapy and the occurrence of breast cancer (Example 5.1). Details on the C.A.MAN-analysis are given in Table 5.7. The nonparametric maximum likelihood estimate consists of 5 groups. A comparison with the maximum likelihood estimate under homogeneity shows that there is a clear evidence for heterogeneity (difference in 2 × log-likelihood about 34). As further analysis shows, there might be fewer

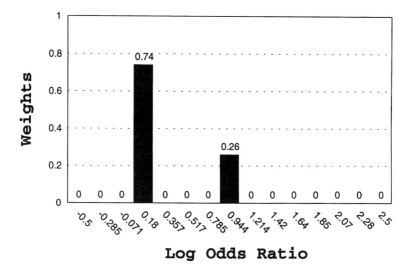

Figure 5.4. NPMLE of heterogeneity distribution of meta-analysis oral contraceptive and breast cancer.

components required than 5. As can be seen in Table 5.7, at least 3 components are required in the mixture. The NPMLE of the heterogeneity distribution can be seen in Figure 5.5.

5.6   Classification of studies using posterior Bayes

Having found the mixing distribution P and its estimate $\hat{P}$ (and thereby an estimate of the heterogeneity involved in the meta-analysis), it is not clear to which component of the subpopulations each study result x belongs, and one might be interested in classifying each study result x into one of the components. Here we denote by x the observed effect measure in the study at hand. This is of particular interest if P is discrete, nonparametric, since in this case the mixture components can be thought of as *disjoint classes* into which the population can be partitioned. Classification can be accomplished by means of the posterior distribution

$$f(\lambda_j | x) = \frac{f(x|\lambda_j)p_j}{f(x, P)} = \frac{f(x|\lambda_j)p_j}{\displaystyle\sum_{l=1}^{k} f(x, \lambda_l)p_l} \qquad (5.10)$$

Parameter	Weights	Log-likelihood
$k = 5$		
−1.0688	0.0341	−16.2211
−0.2874	0.2452	
0.0247	0.2780	
0.1361	0.3009	
0.5550	0.1418	
$k = 4$		
−1.0691	0.0341	−16.5188
−0.2868	0.2517	
0.0783	0.5616	
0.5443	0.1527	
$k = 3$		
−0.3365	0.2804	−17.6306
0.0778	0.5671	
0.5446	0.1524	
$k = 2$		
−0.0869	0.5350	−25.7898
0.1941	0.4650	
$k = 1$		
0.0560	1.0000	−33.1914

Table 5.7. Nonparametric maximum likelihood estimate for meta-analysis on the relationship of hormone replacement therapy and the occurrence of breast cancer

such that x is classified into component j with

$$f(\lambda_j \mid x) = \max_i f(\lambda_i \mid x) \qquad (5.11)$$

with $1 \leq i \leq k$. Again, to put this into practical terms we have to replace P by its maximum likelihood estimator $\hat{P}$ and (5.11) takes the form

$$f(\hat{\lambda}_j \mid x) = \max_i f(\hat{\lambda}_j \mid x) = \max_i \frac{f(x \mid \hat{\lambda}_i)\hat{p}_i}{f(x, \hat{P})}. \qquad (5.12)$$

As we have seen in Table 5.7, at least $k = 3$ components are required to explain the heterogeneity found. Table 5.8 provides the classification of each study into one of the three components. It is now possible to group the studies according to the classifier and look for attributes

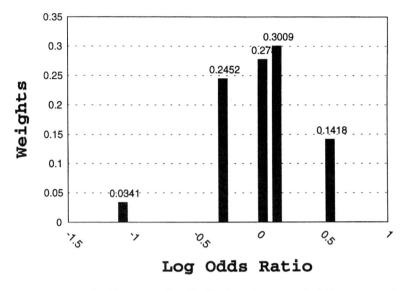

Figure 5.5. NPMLE of heterogeneity distribution of meta-analysis hormone replacement therapy and breast cancer.

which the grouped studies have in common, and thus hopefully explain the heterogeneity found. However, in this case we have no further information on these studies available. To demonstrate this point we consider a further example.

Example 5.9: The nine studies summarized in Table 5.9 gave evidence of the effect of nitrogen dioxide (NO_2) on respiratory disease in children. The studies were all prospective, but the model used for analysis of the data varied from study to study. All studies were adjusted so that the estimated effect was for an increase of about $30\mu g/m^3$ in NO_2 exposure. The effect was summarized as an odds ratio for the increase in respiratory disease in children. The nine studies giving estimates for children in the age range of 6 to 12 years are presented in Table 5.9 (after Hasselblad 1994). C.A.MAN estimates the heterogeneity involved in this meta-analysis to consist of *two* components: a relative risk of 1.0887 receiving weight 0.3836 and an elevated relative risk of 1.2815 receiving weight 0.6164. Classification according to rule (5.12) leads to column 5 in Table 5.9. Evidently, the heterogeneity is explained to a large extent by the method of analysis used in each study. The logistic regression model was used in 6 studies leading to a higher estimate of relative risk. This estimate might also be considered as the more valid estimate of

Study number	Log-odds ratio	Classification
1	0.10436	2
2	−0.03046	2
3	0.76547	2
4	−0.19845	1
5	−0.10536	2
6	−0.11653	2
7	0.09531	2
8	0.26236	2
9	−0.26136	1
10	0.45742	3
11	−0.59784	1
12	−0.35667	2
13	−0.10536	2
14	−0.31471	1
15	−0.10536	2
16	0.02956	2
17	0.60977	3
18	−0.30111	1
19	0.01980	2
20	0.00000	2
21	−0.04082	2
22	0.02956	2
23	0.18232	2
24	0.26236	2
25	0.32208	2
26	0.67803	2
27	−0.96758	1
28	0.91629	3
29	0.32208	2
30	−1.13943	1
31	−0.47804	1
32	0.16551	2
33	0.46373	3
34	−0.52763	1
35	0.10436	2
36	0.55389	3

Table 5.8. C.A.MAN classification of the 36 studies on the relationship of hormone replacement therapy and breast cancer into the three estimated components using the classification rule (5.12)

Author	Model for analysis	Estimated Odds ratio	95% C.I.	Classifier
Melia *et al.* (1977)	multiple logistic	1.31	(1.16, 1.48)	2
Melia *et al.* (1979)	multiple logistic	1.24	(1.09, 1.42)	2
Melia *et al.* (1980)	multiple logistic	1.53	(1.04, 2.24)	2
Melia *et al.* (1982)	multiple logistic	1.11	(0.83, 1.49)	1
Ware *et al.* (1984)	two arm experiment	1.08	(0.97, 1.19)	1
Neas *et al.* (1990)	multiple logistic	1.47	(1.17, 1.86)	2
Ekwo *et al.* (1983)	multiway contingency table	1.10	(0.79, 1.53)	2
Dijkstra *et al.* (1990)	multiple logistic	0.94	(0.66, 1.33)	1
Keller *et al.* (1979)	two arm experiment	1.10	(0.79, 1.54)	2

Table 5.9. Nine studies on the effect of nitrogen dioxide exposure on respiratory disease in children (Hasselblad 1994)

relative risk, since the logistic regression model is capable of adjustment for more potential confounders simultaneously.

5.7   Adjusting for heterogeneity

In this approach interest is shifted from a detailed modeling of the observed heterogeneity to a method where the heterogeneity is adjusted. This can be accomplished by means of a random effects model

$$\hat{\lambda}_i = \lambda_i + \varepsilon_i = \mu + \delta_i + \varepsilon_i, \text{ with } \delta_i = \lambda_i - \mu \qquad (5.13)$$

and $E(\lambda_i) = \mu$. The two error terms δ_i and ε_i correspond to the two random mechanisms involved in the two-stage sampling: in the *first* stage a study with number i is picked from the population of all studies, leading to a subpopulation mean measure λ_i with error δ_i, having mean 0 and variance τ^2. This variance is assumed to be *unknown*. In the *second* stage, given study i with parameter λ_i, a study estimate is found with value $\hat{\lambda}_i$ and associated error ε_i, having mean 0 and known variance σ_i^2. It is assumed that the two random mechanisms are *independent*. Then, $E(\hat{\lambda}_i) = \mu + E(\delta_i) + E(\varepsilon_i) = \mu$ and

$$\mathrm{Var}(\hat{\lambda}_i) = \mathrm{Var}(\delta_i) + \mathrm{Var}(\varepsilon_i) = \tau^2 + \sigma_i^2 = \sigma^{*2}_i \qquad (5.14)$$

What are the consequences of this model? Evidently, the mean structure is unchanged. However, we observe a change in the variance structure, in that an increase of the variance might occur. The changed variance structure is now used to compute a pooled estimator according to (5.15):

$$\hat{\mu}_{\mathrm{DL}} = \frac{\displaystyle\sum_{i=1}^{n} \hat{\lambda}_i/(\hat{\tau}^2 + \sigma_i^2)}{\displaystyle\sum_{i=1}^{n} 1/(\hat{\tau}^2 + \sigma_i^2)} \qquad (5.15)$$

Evidently, an estimator of τ^2 is required and has been suggested by DerSimonian and Laird (1986). This estimator is given as $\hat{\tau}^2 = (\chi^2_{(n-1)} - (n-1))/(\Sigma_i w_i - \Sigma_i w_i^2/\Sigma_i w_i))^*$ with $w_i = 1/\sigma_i^2$ for $i = 1, ..., n$ and is discussed in Section 6.2 in detail. Here, we just want to mention *two* extreme cases. The first is when $\hat{\tau}^2 = 0$. Then, there is *no* difference from the conventionally pooled one (5.2). If $\hat{\tau}^2$ is large relative to the σ_i^2, then $\hat{\mu}_{\mathrm{DL}} = (1/n)\sum_{i=1}^{n} \hat{\lambda}_i$, the arithmetic mean of the observed effect measures. Quite similar to the homogeneous version, its variance is given as

$$\mathrm{Var}(\hat{\mu}_{\mathrm{DL}}) = \frac{1}{\displaystyle\sum_{i=1}^{n} 1/(\hat{\tau}^2 + \sigma_i^2)}$$

Since $1/\sum_{i=1}^{n} 1/(\hat{\tau}^2 + \sigma_i^2) \geq 1/\sum_{i=1}^{n} 1/\sigma_i^2$, adjusting for a random effect in the described way will *always* lead to a confidence interval which is *at least* as large as the one based upon the conventional estimator.

---

* If $\hat{\tau}^2 < 0$ it is truncated at 0.

Example 5.10: We continue our discussion of Example 5.1. We find $\hat{\lambda}_{best}$ as 0.05598 with 95% C.I. [0.0118, 0.1002]. Retransformation to the relative risk scale provides $\hat{\psi}_{best} = \exp(\hat{\lambda}_{best}) = 1.05758$ with 95% C.I. [1.01187, 1.10535]. In the random effects model we find $\hat{\mu}_{DL} = 0.0337$ with 95% C.I. [–0.0627, 0.1302], or on the relative risk scale $\hat{\psi}_{DL} = 1.03432$ with associated 95% C.I. [0.9392, 1.1390] indicating that the small significant effect mentioned before has vanished. The corresponding effect models are visualized in Figure 5.6.

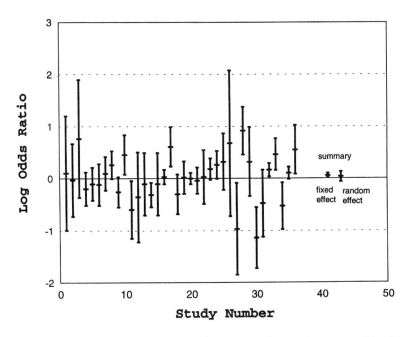

Figure 5.6. Effect estimates (log-odds ratios) of 36 studies on the relationship of hormone replacement therapy and breast cancer with summary measures based upon fixed effects and random effects model.

Frequently, there are objections to this approach. It is argued that — in the presence of heterogeneity — there is no common ground for computing a summary measure. If one follows the opinion that only for a homogeneous population may a summary measure be validly computable, then this objection is substantiated. However, one could also compute a summary measure in the presence of heterogeneity, though the *interpretation* would be different. It would provide an estimate of the *overall mean* effect measure in the population, knowing that this might be different in subparts of the population. In contrast, the overall mean measure of effect would be also estimating the effect in subsets of an *homogeneous* population.

CHAPTER 6

Moment estimators of the variance of the mixing distribution

6.1   A variance decomposition with a latent factor

In a variety of applications the situation of extra-population heterogeneity occurs. In particular, this is the case if there is good reason to model the variable of interest X through a density of parametric form $f(x, \lambda)$ with a scalar parameter λ. For a given subpopulation, the density $f(x, \lambda)$ might be very suitable, but the value of λ is not able to cover the whole population of interest. See Figure 6.1. In these situations we speak of *extra hetero-geneity* which might be caused by *unobserved covariates* or *clustered obser-vations* such as herd clustering in estimating animal infection rates. An introductory discussion can be found in Aitkin *et al.* (1990, p. 213) and the references given there; see also the review of Pendergast *et al.* (1996).

In this chapter it is understood that *extra-population heterogeneity,* or in brief *population heterogeneity,* refers to the situation that the parameter of interest, λ, varies in the population and sampling has not taken this into account (e.g., it has not been observed from which subpopulation (defined by the values of λ) the datum is coming from). As will be clear from equation (6.1) below, inference is affected by the occurrence of extra-population heterogeneity.

For example, variances of estimators of interest are often largely increased, leading to wider confidence intervals as compared to conventional ones. The adjustment of these variances requires the estimation of the variance of the distribution associated with the extra-heterogeneity. The main purpose of this chapter is to present moment estimators for the heterogeneity variance in a simple manner.

To be more precise, if λ is itself varying with distribution P and associated density $p(\lambda)$, the (unconditional) marginal density of X is $f(x) = f(x, P) = \int_{-\infty}^{\infty} f(x, \lambda)p(\lambda)d\lambda$. We are interested in the separation of the (unconditional) variance into *two* terms:

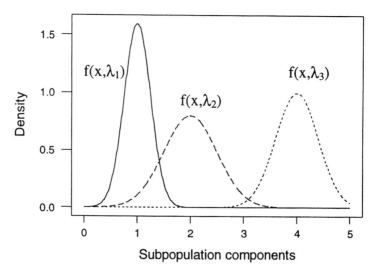

Figure 6.1. The occurrence of population heterogeneity.

$$\mathrm{Var}(X) = \int_{-\infty}^{\infty} \mathrm{Var}(X|\lambda)p(\lambda)d\lambda + \int_{-\infty}^{\infty} (\mu(\lambda) - \mu_x)^2 p(\lambda)d\lambda$$

$$= E(\sigma^2(\lambda)) + \delta^2$$

(6.1)

where $\mu(\lambda)$ is the conditional expected value $E(X|\lambda)$ and μ_x is the overall mean of X. Of course, $\sigma^2(\lambda) = \mathrm{Var}(X|\lambda)$. In some situations which we will demonstrate later,

$$\delta^2 = \int_{-\infty}^{\infty} (\mu(\lambda) - \mu_x)^2 p(\lambda)d\lambda$$

is of the form

$$\delta^2 = \gamma \int_{-\infty}^{\infty} (\lambda - \mu_\lambda)^2 p(\lambda)d\lambda = \gamma\tau^2$$

(6.2)

where $\mu_\lambda = \mu$ is now the mean* of λ and $\tau^2 = \int_{-\infty}^{\infty} (\lambda - \mu_\lambda)^2 p(\lambda)d\lambda$.

* In the succeeding text, we will use μ as the mean of λ.

One can easily identify here that τ^2 is the variance of λ. Thus, in these instances, we can say that (6.1) is a partitioning of the variance due to the *variation in the subpopulation with parameter value* λ and due to the *variance in the heterogeneity distribution P of* λ.

The heterogeneity distribution itself might be continuous (Figure 6.2) or discrete (Figure 6.3). Its structure is *not* the main interest, which is rather a certain functional, its variance. One can also think of (6.1) as an analysis-of-variance partition with a *latent factor* having distribution P.

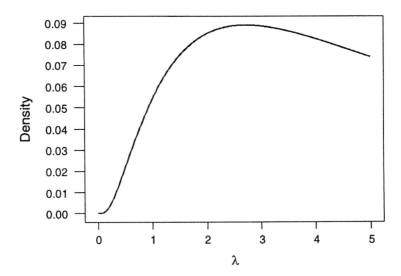

Figure 6.2. Distribution of heterogeneity parameter λ having mean μ and variance τ^2.

Example 6.1: We consider $f(x_i, \lambda, \sigma_i^2) = \varphi((x_i - \lambda)/\sigma_i)/\sigma_i$, where $\varphi(z) = \exp(-z^2/2)/\sqrt{2\pi}$, in other words, the normal distribution with sample unit specific variance σ_i^2 is assumed to be known. Then, according to (6.1)

$$\mathrm{Var}(X_i) = \int_{-\infty}^{\infty} \mathrm{Var}(X_i|\lambda)p(\lambda)d\lambda + \int_{-\infty}^{\infty} (\mu(\lambda) - \mu_x)^2 p(\lambda)d\lambda = \sigma_i^2 + \tau^2.$$

Example 6.2: We consider $f(x, \lambda) = \mathrm{Po}(x, \lambda)$. Then, (6.1) leads to

$$\mathrm{Var}(X) = \int_{-\infty}^{\infty} \lambda p(\lambda)d\lambda + \tau^2 = \mu + \tau^2.$$

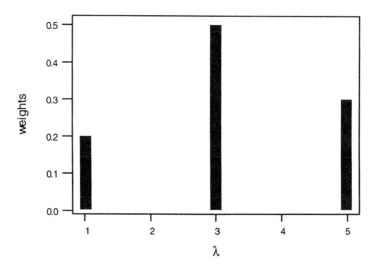

Figure 6.3. Distribution of heterogeneity parameter λ having mean μ and variance τ^2.

Similarly, let $f(x, \lambda, m) = \binom{m}{x} \lambda^x (1 - \lambda)^{m-x}$. (6.1) takes on the form

$$\mathrm{Var}(X) = \int_{-\infty}^{\infty} m\lambda(1-\lambda)p(\lambda)d\lambda + m^2\tau^2 = m\int_{-\infty}^{\infty} \lambda - \lambda^2 p(\lambda)d\lambda + m^2\tau^2$$

$$= m[\mu - (\tau^2 + \mu^2)] + m^2\tau^2 = m\mu(1-\mu) + m(m-1)\tau^2.$$

In both cases, if $\tau^2 = 0$ we achieve the conventional Poisson and binomial variances, respectively.

Example 6.3: We consider the exponential density $f(x, \lambda) = (1/\lambda)e^{-x/\lambda}$ with $E(X|\lambda) = \lambda$ and $\mathrm{Var}(X|\lambda) = \lambda^2$. According to (6.1),

$$\mathrm{Var}(X) = \int_{-\infty}^{\infty} \lambda^2 p(\lambda)d\lambda + \tau^2 = \tau^2 + \mu^2 + \tau^2 = 2\tau^2 + \mu^2.$$

We now present the proof of the decomposition (6.1).

Theorem 6.1: Let X have conditional variance $\mathrm{Var}(X|\lambda)$, conditional mean $\mu(\lambda)$ and unconditional variance $\mathrm{Var}(X)$, and unconditional mean μ_x. Then

$$\mathrm{Var}(X) \;=\; \int_{-\infty}^{\infty} \mathrm{Var}(X|\lambda) p(\lambda) d\lambda + \int_{-\infty}^{\infty} (\mu(\lambda) - \mu_x)^2 p(\lambda) d\lambda \;.$$

Proof. Let $f(x) = \int_{-\infty}^{\infty} f(x|\lambda) p(\lambda) d\lambda$ be the (unconditional) marginal density function of a random variable X. Then,

$$\mathrm{Var}(X) \;=\; E(X - E(X))^2 \;=\; \int_{-\infty}^{+\infty} f(x)(x - E(X))^2 dx \;,$$

where

$$E(X) \;=\; \int_{-\infty}^{+\infty} f(x)x \; dx \;=\; \int_{-\infty}^{\infty} x \int_{-\infty}^{\infty} f(x, \lambda) p(\lambda) d\lambda \; dx \;=\; \int_{-\infty}^{\infty} \mu(\lambda) p(\lambda) d\lambda \;.$$

Here, $\mu(\lambda)$ is the conditional expected value of X: $\mu(\lambda) = E(X|\lambda) = \int_{-\infty}^{+\infty} x \, f(x|\lambda) \, dx$. The unconditional expected value of X is denoted by μ_x. Note that E is the statistical operator of taking the expected value. $\mathrm{Var}(X)$ can be written further as

$$\mathrm{Var}(X) \;=\; \int_{-\infty}^{\infty} (x - \mu_x)^2 f(x) dx \;=\; \int_{-\infty}^{\infty} (x - \mu_x)^2 \int_{-\infty}^{\infty} f(x|\lambda) p(\lambda) d\lambda \; dx$$

$$=\; \int_{-\infty}^{\infty} \int_{-\infty}^{\infty} f(x|\lambda)(x - \mu(\lambda) + \mu(\lambda) - \mu_x)^2 dx \; p(\lambda) d\lambda$$

$$=\; \int_{-\infty}^{\infty} \int_{-\infty}^{\infty} (x - \mu(\lambda))^2 f(x|\lambda) dx \; p(\lambda) d\lambda + \int_{-\infty}^{\infty} (\mu(\lambda) - \mu_x)^2 \int_{-\infty}^{\infty} f(x|\lambda) dx \; p(\lambda) d\lambda$$

$$=\; \int_{-\infty}^{\infty} \mathrm{Var}(X|\lambda) p(\lambda) d\lambda + \int_{-\infty}^{\infty} (\mu(\lambda) - \mu_x)^2 p(\lambda) d\lambda$$

which ends the proof.

The intention is to find an estimate of τ^2 without implying knowledge or estimating the latent heterogeneity distribution P. We will first consider the case of the normal distribution with known variance; this is the DerSimonian–Laird approach. Then we will look at members of the exponential family and develop estimators based on the variance decomposition.

As throughout this book, it is assumed that $X_1, \ldots, X_n$ represents a random sample from a population with density $f(x, P)$.

6.2   The DerSimonian–Laird estimator

Consider the situation of example 6.1: $f(x_i, \lambda, \sigma_i^2) = \varphi((x_i - \lambda)/\sigma_i)/\sigma_i$, where φ is standard normal density. This setting arises typically in *meta-analysis* where x_i represents the effect measure coming from study i and σ_i^2 is the associated *variance*. This situation has been studied in detail in Chapter 5. We would like to point out that σ_i^2 is the *conditional variance given the datum x_i arising from the component with valid parameter λ*. DerSimonian and Laird consider the statistic

$$\chi^2 = \sum_{i=1}^{n} (X_i - \mu)^2 / \sigma_i^2 \qquad (6.3)$$

which is approximately χ_n^2-distributed with n degrees of freedom under homogeneity. However, μ is *unknown* and it is straightforward to replace it by an efficient estimator $\hat{\mu} = \sum_{i=1}^{n} w_i X_i / \sum_{i=1}^{n} w_i$, where $w_i = 1/\sigma_i^2$. This estimator* is efficient in the sense that it minimizes the variance of all linear estimators of the form $\sum_{i=1}^{n} p_i X_i$, with $p_i \geq 0$ for all $i - 1, ..., n$ and $\sum_{i=1}^{n} p_i - 1$. Then, (6.3) becomes

$$\chi^2 = \sum_{i=1}^{n} (X_i - \hat{\mu})^2 / \sigma_i^2 \qquad (6.4)$$

which is now χ_{n-1}^2 with $n - 1$ degrees of freedom under homogeneity. DerSimonian and Laird (1986) equate the expected value of χ^2 to its sample version and solve the moment equation for τ^2.

Lemma 6.1 Let χ^2 be defined as in (6.4). Then,

$$E(\chi^2) = n - 1 + \tau^2(\Sigma_i w_i - \Sigma_i w_i^2 / \Sigma_i w_i) \qquad (6.5)$$

Proof. We write $(X_i - \hat{\mu})^2$ as $(X_i - \mu + \mu - \hat{\mu})^2$ and consider three terms:

(a) $(X_i - \mu)^2$ which has expected value $\sigma_i^2 + \tau^2$,

(b) $2(X_i - \mu)(\mu - \Sigma_j w_j X_j / \Sigma_j w_j) = -2\Sigma_j w_j (X_i - \mu)(X_j - \mu)/\Sigma_j w_j$ which has expected value $-2w_i(\sigma_i^2 + \tau^2)/\Sigma_j w_j = -2/\Sigma_j w_j - 2\tau^2 w_i/\Sigma_j w_j$

(c) $(\mu - \Sigma_j w_j X_j / \Sigma_j w_j)^2 = (1/\Sigma_j w_j)^2 \, \Sigma_j w_j^2 (x_j - \mu)^2$ which has expected value $(1/\Sigma_j w_j)^2 \, \Sigma_j w_j^2 (\sigma_j^2 + \tau^2) = 1/\Sigma_j w_j + \tau^2 \, \Sigma_j w_j^2/(\Sigma_j w_j)^2$.

---

* This estimator is also sometimes called the *Mantel–Haenszel-estimator.*

Taking the sum of all three terms gives

$$E(X_i - \hat{\mu})^2 / \sigma_i^2 = 1 + \tau^2 w_i - 2w_i/\Sigma_j w_j - 2\tau^2 w_i^2/\Sigma_j w_j$$
$$+ w_i/\Sigma_j w_j + \tau^2 \ w_i \ \Sigma_j w_j^2/(\Sigma_j w_j)^2$$

and summing over all i

$$E(\chi^2) = n + \tau^2 \ \Sigma_i w_i - 2 - 2\tau^2 \ \Sigma_i w_i^2/\Sigma_j w_j$$
$$+ 1 + \tau^2 \ \Sigma_j w_j^2/\Sigma_j w_j = n - 1 + \tau^2(\Sigma_i w_i - \Sigma_i w_i^2/\Sigma_i w_i)$$

which completes the proof. As a consequence we have the following theorem.

Theorem 6.2 (DerSimonian and Laird 1986):

Let χ^2 be defined as in (6.4). Then, the *moment estimator* of τ^2 is given by

$$\hat{\tau^2} = [\chi^2 - (n-1)]/ \ (\Sigma_i w_i - \Sigma_i w_i^2/\Sigma_i w_i) \qquad (6.6)$$

In case $\chi^2 < n - 1$, the estimator is *truncated* to zero.

If there is homogeneity, χ^2 will be close to $n - 1$ ($E(\chi^2) = n - 1$) and $\hat{\tau^2}$ near 0. If there is more heterogeneity, χ^2 will be larger than $n - 1$ and $\hat{\tau^2}$ larger than 0. Note that by construction the estimator (6.6) is *unbiased*. The consequences for computing a *pooled estimator* are as follows: one replaces the individual variances by the more appropriate ones in the weights: $\omega_i = 1/(\sigma_i^2 + \hat{\tau^2})$. This leads to the *pooled* estimator

$$\hat{\mu}_{DS} = \Sigma_i X_i \omega_i / \Sigma_i \omega_i \qquad (6.7)$$

which has variance

$$\mathrm{Var}(\Sigma_i X_i \omega_i / \Sigma_i \omega_i) = 1/\Sigma_i \omega_i .$$

It is natural to compare the variance of $\hat{\mu}_{DS}$ with $\hat{\mu}$.

Theorem 6.3: The following inequality holds:

$$\mathrm{Var}(\hat{\mu}_{DS}) \geq \mathrm{Var}(\hat{\mu}) \qquad (6.8)$$

Proof. We have $\sigma_i^2 + \hat{\tau}^2 \geq \sigma_i^2$, or $\omega_i \leq w_i$ for all $i = 1, ..., n$. Consequently, $\Sigma_i \omega_i \leq \Sigma_i w_i$, or $1/\Sigma_i \omega_i \geq 1/\Sigma_i w_i$, which is the desired result.

If $\hat{\tau}^2 = 0$, $\hat{\mu}_{DS} = \hat{\mu}$. If $\hat{\tau}^2$ is large in comparison to $\max\{\sigma_1^2, ..., \sigma_n^2\}$, then $\hat{\mu}_{DS}$ will be close to the arithmetic mean $\Sigma_i X_i/n$.

Example 6.4: We consider the following meta-analytic application. The studies summarized in Table 6.1 gave evidence of the effect of nitrogen dioxide (NO_2) on respiratory disease in children. The studies were all prospective, but the model used for analysis of the data varied from study to study. All studies were adjusted so that the estimated effect was for an increase of about $30\mu g/m^3$ in NO_2 exposure. The effect was summarized as an odds ratio for the increase in respiratory disease in children. The nine studies giving estimates for children in the age range of 6 to 12 years are given in the following Table 6.1 (after Hasselblad 1994).

Author	Model for analysis	Estimated odds ratio	95% C.I.
Melia *et al.* (1977)	multiple logistic	1.31	(1.16, 1.48)
Melia *et al.* (1979)	multiple logistic	1.24	(1.09, 1.42)
Melia *et al.* (1980)	multiple logistic	1.53	(1.04, 2.24)
Melia *et al.* (1982)	multiple logistic	1.11	(0.83, 1.49)
Ware *et al.* (1984)	two arm experiment	1.08	(0.97, 1.19)
Neas *et al.* (1990)	multiple logistic	1.47	(1.17, 1.86)
Ekwo *et al.* (1983)	multiway contingency table	1.10	(0.79, 1.53)
Dijkstra *et al.* (1990)	multiple logistic	0.94	(0.66, 1.33)
Keller *et al.* (1979)	two arm experiment	1.10	(0.79, 1.54)

Table 6.1. Nine studies on the effect of NO_2 exposure on respiratory disease in children

The data given in Table 6.1 are not yet in a form which would allow the analysis of type (6.4) or (6.7). The individual variances have to be extracted from the given confidence intervals. Conditional on the assumption that these intervals are constructed on the log-scale using the formula

$$\lambda_{L, R} = \log \text{ odds ratio} \pm 1.96 \text{ SE}$$

we can solve for the SE as

$$SE = [\lambda_R - \lambda_L]/ (2 \times 1.96) \qquad (6.9)$$

and we get the variance by squaring (6.9). Table 6.2 shows the log-odds ratios for the nine studies with associated variance.

Study Number	Author	Log-odds ratio	Variance
1	Melia *et al.* (1977)	0.270027	0.0038624
2	Melia *et al.* (1979)	0.215111	0.0045521
3	Melia *et al.* (1980)	0.425268	0.0383096
4	Melia *et al.* (1982)	0.104360	0.0222790
5	Ware *et al.* (1984)	0.076961	0.0027192
6	Neas *et al.* (1990)	0.385262	0.0139850
7	Ekwo *et al.* (1983)	0.095310	0.0284327
8	Dijkstra *et al.* (1990)	−0.061875	0.0319510
9	Keller *et al.* (1979)	0.095310	0.0289959

Table 6.2. Nine studies on the effect of NO_2 exposure on respiratory disease in children: log-odds ratio and variance

It is now possible to provide an heterogeneity analysis. The χ^2-value according to (6.4) is 13.4376 with a P-value of 0.0977 (8 df). The pooled estimator is $\hat{\mu} = 0.1776$ with standard error $(1/\Sigma_i w_i)^{1/2} = 0.03029$. Incorporation of heterogeneity leads to an τ^2 estimate of 0.0064 and $\hat{\mu}_{DS} = 0.1851$ with standard error $(1/\Sigma_i \omega_i)^{1/2} = 0.04528$. As expected the standard error of the DerSimonian–Laird estimator is larger than the conventionally pooled one. Figure 6.4 provides a graphical analysis of the situation.

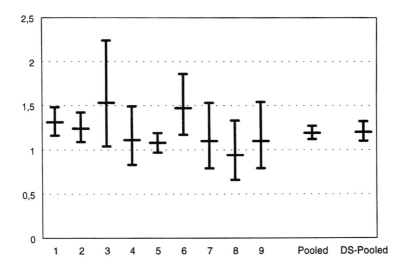

Figure 6.4. Nine studies on the effect of NO_2 exposure on respiratory disease in children with 95% confidence interval; pooled and DerSimonian–Laird pooled estimator of effect with 95% confidence interval.

Whether the conventionally pooled or the pooled estimator adjusting for heterogeneity is used, in both cases we find the 95% confidence interval to *not* include the baseline risk of 1.

6.3 The Böhning–Sarol estimator

We consider again the basic variance decomposition (6.1): $\text{Var}(X) = E(\sigma^2(\lambda)) + \delta^2 = E(\sigma^2(\lambda)) + \gamma\tau^2$. If it is possible to solve the integral $E(\sigma^2(\lambda))$, it should be quite possible to give an estimator for τ^2. The idea is very simple: we replace $\text{Var}(X)$ and $E(\sigma^2(\lambda))$ on the right-hand side of (6.1) by their respective sample estimates and obtain an estimate for τ^2. If $\hat{\tau}^2$ is negative, we truncate it to 0 (this is very similar to the estimator proposed by DerSimonian and Laird in Section 6.2).

Example 6.5: We consider $f(x_i, \lambda, \sigma_i^2) = \varphi((x_i - \lambda)/\sigma_i)/\sigma_i$, where $\varphi(z) = \exp(-z^2/2)/\sqrt{2\pi}$, in other words, the normal distribution with sample unit specific variance σ_i^2 is assumed to be known. Then, according to Example 6.1 we have that $\text{Var}(X_i) = \sigma_i^2 + \tau^2$. Suppose for the time being $\sigma_i^2 = \sigma^2$ for all $i = 1, ..., n$. Then τ^2 can be estimated as $\hat{\tau}^2 = S^2 - \sigma^2$. Note that $E(\hat{\tau}^2) = E(S^2) - \sigma^2 = \text{Var}(X_i) - \sigma^2 = \tau^2$. Here $S^2 = 1/(n-1)\Sigma_i (X_i - \overline{X})^2$ and $\overline{X} = (X_1 + ... + X_n)/n$.

Example 6.6: We consider $f(x, \lambda) = \text{Po}(x, \lambda)$. According to Example 6.2, $\text{Var}(X) = \mu + \tau^2$ and $\hat{\tau}^2 = S^2 - \overline{X}$. This quantity has also been referred to as a measure of Poisson overdispersion (Böhning 1994). Note that $E(\hat{\tau}^2) = \tau^2$.

Similarly, for the binomial $f(x, \lambda, m) = \binom{m}{x} \lambda^x(1-\lambda)^{m-x}$ we have that

$$\text{Var}(X) = m\mu(1-\mu) + m(m-1)\tau^2.$$

Replacing again μ by $\overline{X}/m$ and $\text{Var}(X)$ by $S^2 = 1/(n-1)\Sigma_i (X_i - \overline{X})^2$ we are led to

$$\hat{\tau}^2 = \frac{[S^2 - \overline{X}(1 - \overline{X}/m)]}{m(m-1)}. \tag{6.10}$$

The estimator (6.10) is *not* unbiased. We look at the 3 terms involved in (6.10): $S^2, \overline{X}, \overline{X}^2$. $E(S^2) = m\mu(1-\mu) + m(m-1)\tau^2$, $E(\overline{X}) = m\mu$, and $E(\overline{X}^2) = \text{Var}(\overline{X}) + (E\overline{X})^2 = [m\mu(1-\mu) + m(m-1)\tau^2]/n + m^2\mu^2$. Putting these terms together gives

$$E(S^2) - E(\overline{X}) + E(\overline{X^2})/m$$

$$= m\mu(1 - \mu) + m(m - 1)\tau^2 - m\mu + [m\mu(1 - \mu) + m(m - 1)\tau^2]/nm + m\mu^2$$

$$= m(m - 1)\tau^2 + [m\mu(1 - \mu) + m(m - 1)\tau^2]/nm$$

and finally

$$E(\hat{\tau}^2) = \tau^2 + \mathrm{Var}(X)/[nm^2(m - 1)]$$

This bias is practically negligible. For example, if $n = m = 10$, then the bias is of order 10^{-4}. If $n = 100$, $m = 10$, the bias is of order 10^{-5}. Also, the bias becomes smaller if n, m, or both increase. A bias-corrected estimate of τ^2 is given by

$$\hat{\tau}^2_{corrected} = \frac{nm - 1}{m^2(m - 1)n}S^2 - \frac{\overline{X}}{m}\left(1 - \frac{\overline{X}}{m}\right)/(m - 1) \qquad (6.11)$$

Theorem 6.4: The estimator (6.11) is unbiased.

Proof.

$$E(\hat{\tau}^2_{corrected}) = \frac{nm - 1}{m^2(m - 1)n}(m\mu(1 - \mu) + m(m - 1)\tau^2) - \mu/(m - 1)$$

$$+ E(\overline{X^2})/(m^2(m - 1)) = \frac{nm - 1}{mn}(\mu(1 - \mu)/(m - 1) + \tau^2) - \mu/(m - 1)$$

$$+ \frac{m\mu(1 - \mu) + m(m - 1)\tau^2}{nm^2(m - 1)} + \frac{m^2\mu^2}{m^2(m - 1)} = \tau^2$$

Example 6.7: We consider again the exponential density $f(x, \lambda) = (1/\lambda)e^{-x/\lambda}$ with $E(X|\lambda) = \lambda$ and $\mathrm{Var}(X|\lambda) = \lambda^2$. According to Example 6.3, we have that $\mathrm{Var}(X) = 2\tau^2 + \mu^2$, leading to $\hat{\tau}^2 = (S^2 - \overline{X^2})/2$. Due to the fact that $E(\overline{X^2}) \neq (E\overline{X})^2$ the estimator $\hat{\tau}^2$ is *not* unbiased. A direct computation shows that

$$E(\hat{\tau}^2) = \tau^2 - \frac{1}{n}(\tau^2 + \mu^2/2).$$

We come back to Example 6.5 and allow the variances σ_i^2 to be different. This increases the complexity considerably and a careful analysis of the situation is required. We consider the natural extension of $\hat{\tau}^2 = S^2 - \sigma^2$, namely,

$$T = \frac{1}{n-1}\Sigma_i(X_i - \hat{\mu})^2 - \frac{1}{n}\Sigma_i\sigma_i^2 \tag{6.12}$$

where $\hat{\mu}$ is the pooled estimator $\Sigma_i\, w_i X_i/\Sigma_i w_i$ with $w_i = 1/\sigma_i^2$.

Theorem 6.5: For the statistic T defined in (6.12) we have:

$$E(T) = \left[\frac{n-2}{n-1} + \frac{n}{n-1}\frac{\Sigma_i w_i^2}{(\Sigma_i w_i)^2}\right]\tau^2 + \left[\frac{1}{n-1} - \frac{1}{n}\right]\Sigma_i\frac{1}{w_i} - \frac{n}{n-1}\frac{1}{\Sigma_i w_i}$$

Proof. We write $(1/(n-1))\Sigma_i\,(X_i - \hat{\mu})^2$ as

$$\frac{1}{n-1}\Sigma_i(X_i - \mu)^2 - 2\frac{1}{n-1}\Sigma_i(x_i - \mu)(\hat{\mu} - \mu) + \frac{n}{n-1}(\hat{\mu} - \mu)^2 .$$

The three terms involved in this sum have the following expected values:

$$E\left(\frac{1}{n-1}\Sigma_i(X_i - \mu)^2\right) = \frac{1}{n-1}\Sigma_i\sigma_i^2 + \frac{n}{n-1}\tau^2$$

$$E\left(2\frac{1}{n-1}\Sigma_i(X_i - \mu)(\hat{\mu} - \mu)\right) = -\frac{2n}{n-1}\frac{1}{\Sigma_i w_i} - \frac{2}{n-1}\tau^2$$

$$E\left\{\frac{1}{n-1}\frac{1}{(\Sigma_i w_i)^2}(\Sigma_i w_i(X_i - \mu))^2\right\} = \frac{n}{n-1}\frac{1}{\Sigma_i w_i} + \frac{n}{n-1}\frac{\Sigma_i w_i^2}{(\Sigma_i w_i)^2}\tau^2$$

Summing up these three values and subtracting $(1/n)\Sigma_i\,\sigma_i^2$ leads to the result and finishes the proof.

It is clear how an unbiased estimator can be constructed from T: let

$$\hat{\tau}^2_{corrected} = (T - \alpha)/\beta \tag{6.13}$$

where $\alpha = [1/(n-1) - 1/n]\Sigma_i\, 1/w_i - (n/(n-1))(1/(\Sigma_i w_i))$ and $\beta = (n-2)/(n-1) + (n/(n-1))\Sigma_i w_i^2/\Sigma_i w_i^2$, and $\hat{\tau}^2_{corrected}$ is *unbiased* now for τ^2. It might be of interest to compare the estimator $(T - \alpha)/\beta$ with the DerSimonian–Laird estimator (6.6). If all variances are identical ($\sigma_i^2 = \sigma^2$ for all $i = 1, ..., n$), then

$$[\chi^2 - (n-1)]/(\Sigma_i w_i - \Sigma_i w_i^2/\Sigma_i w_i)$$

$$= \left[\frac{1}{\sigma^2}\Sigma_i(X_i - \bar{X})^2 - (n-1)\right]/[n/\sigma^2 - (n/\sigma^4)/(n/\sigma^2)]$$

$$= \frac{1}{n-1}\Sigma_i(X_i - \bar{X})^2 - \sigma^2$$

which coincides with (6.13) in this case. In general, they are *not identical* and it might be of interest to compare their efficiencies. Table 6.3 shows the bootstrapped standard errors of the estimators (6.6) and (6.13) for the data of Example 6.4. It seems that (6.6) is more efficient.

Estimator	Standard error
(6.6)	0.00486
(6.13)	0.00701

Table 6.3. Bootstrapped standard errors of estimators (6.6) and (6.13) for data of Example 6.4

In the next section we will consider generalizations of this idea.

6.4 Estimation of a binomial or Poisson rate under heterogeneity

Standardized Mortality Ratio. The SMR is the ratio of an observed number of mortality cases O and an expected (non-random) number of mortality cases e. Frequently the assumption of a Poisson distribution for O is used with mean $E(O \mid e, \lambda) = \mu(\lambda) = \lambda e$. If we allow extra-Poisson variation, the partition of variance is

$$\mathrm{Var}(O) = \int_{-\infty}^{\infty} \mathrm{Var}(O \mid \lambda) p(\lambda) d\lambda + \int_{-\infty}^{\infty} (\lambda e - \mu e)^2 p(\lambda) d\lambda$$

$$= E(\sigma^2(\lambda)) + e^2\tau^2 = e\mu + e^2\tau^2.$$

Consequently, $\tau^2 = \text{Var}(O)/e^2 - \mu/e$. Here, τ^2 represents the variance of the λs, the theoretical SMRs.

The estimation is complicated here by the fact that the expected mortality cases might differ in a sample. Let O_1, ..., O_n be a random sample of mortality cases with associated expected mortality cases e_1, ..., e_n. Often the sample represents a collection of SMR-values for a set of geographic regions and one is interested in analyzing geographic variation. Let

$$\hat{\tau^2} = \frac{1}{n}\left[\Sigma_i(O_i - e_i\mu)^2/e_i^2 - \mu\Sigma_i\frac{1}{e_i}\right] \qquad (6.14)$$

Evidently, $\hat{\tau^2}$ is an unbiased estimator of τ^2. We consider two unbiased estimates of μ, leading to different formula for obtaining an unbiased estimator of τ^2. The first is the simple (unweighted) mean of the SMRs, that is, $\hat{\mu} = (1/n)\Sigma_i O_i/e_i$. This leads to

$$\hat{\tau^2} = \frac{1}{n-1}\Sigma_i(O_i - e_i\hat{\mu})^2/e_i^2 - \frac{1}{n}\hat{\mu}\Sigma_i\frac{1}{e_i} \qquad (6.15)$$

$$= \frac{1}{n-1}\Sigma_i(\text{SMR}_i - \hat{\mu})^2 - \frac{1}{n}\hat{\mu}\Sigma_i\frac{1}{e_i}$$

where we have adjusted for the estimation of μ. We show the following result.

Theorem 6.6: The estimator (6.15) is unbiased for τ^2.

Proof. To find the expected value of (6.15) we write $(O_i - e_i\hat{\mu})^2$ as $(O_i - e_i\mu + e_i\mu - e_i\hat{\mu})^2$ and compute the expected values of the 3 terms:

(a) $E(O_i - e_i\mu)^2 = e_i\mu + e_i^2\tau^2$

(b) $E2(O_i - e_i\mu)(e_i\mu - e_i\hat{\mu}) = -(2/n)\mu e_i - (2/n)\tau^2 e_i^2$

(c) $E(e_i\mu - e_i\hat{\mu})^2 = e_i^2/n^2(\Sigma_j 1/e_j + n\tau^2)$

Dividing by e_i^2, summing up all terms and dividing by $(n-1)$ gives

$$\tau^2(n - 2 + 1)/(n - 1) + \mu\,\Sigma_i\,1/e_i\,(1 - 2/n + 1/n)/(n - 1)$$
$$= \tau^2 + \mu\,\Sigma_i\,1/e_i\frac{1}{n}$$

Therefore,

$$E(\hat{\tau}^2) = E\left[\frac{1}{n-1}\Sigma_i(O_i - e_i\hat{\mu})^2/e_i^2 - \frac{1}{n}\hat{\mu}\Sigma_i\frac{1}{e_i}\right]$$

$$= \tau^2 + \mu\frac{1}{n}\Sigma_i\frac{1}{e_i} - E(\hat{\mu})\frac{1}{n}\Sigma_i\frac{1}{e_i} = \tau^2.$$

This completes the proof.

Alternatively, one could replace μ in (6.14) with a different unbiased estimate of μ, namely, the pooled estimator $\hat{\mu} = \Sigma_iO_i/\Sigma_ie_i$. We first let $\hat{\tau}^2 = (1/(n-1))\Sigma_i(\text{SMR}_i - \hat{\mu})^2 - (1/n)\hat{\mu}\,\Sigma_i(1/e_i)$. This, however, results in a *biased* estimate of τ^2, as is shown in the following theorem.

Theorem 6.7: For the estimator $\hat{\tau}^2 = (1/(n-1))\Sigma_i(\text{SMR}_i - \hat{\mu})^2 - (1/n)\hat{\mu}\,\Sigma_i$ $(1/e_i)$ with $\hat{\mu} = \Sigma_iO_i/\Sigma_ie_i$, we have

$$E(\hat{\tau}^2) = \alpha\mu + \beta\tau^2 \tag{6.16}$$

with $\alpha = 1/(n(n-1))\Sigma_i\,1/e_i - (n/(n-1)/\Sigma_ie_i$ and $\beta = (n-2+n\Sigma_ie_i^2/(\Sigma_ie_i)^2)/(n-1)$.

Proof. To find the expected value (6.16) we write $(O_i - e_i\,\hat{\mu})^2$ as $(O_i - e_i\mu$ $+\, e_i\mu - e_i\hat{\mu})^2$ and compute the expected values of the 3 terms:

(a) $E(O_i - e_i\mu)^2 = e_i\mu + e_i^2\tau^2$

(b) $E\;2(O_i - e_i\mu)(e_i\mu - e_i\hat{\mu}) = -\,2e_i/\Sigma_je_j\,(\mu e_i - \tau^2e_i^2)$

(c) $E(e_i\mu - e_i\hat{\mu})^2 = e_i^2/(\Sigma_je_j)^2\Sigma_j\,(e_j\mu + e_i^2\tau^2)$

Dividing by e_i^2, summing up all terms and dividing by $(n-1)$ gives

$$\mu(\Sigma_i1/e_i - 2n/\Sigma_ie_i + n/\Sigma_ie_i)/(n-1) \tag{6.17}$$

$$+\,\tau^2\left(n-2+n\frac{\Sigma_ie_i^2}{(\Sigma_ie_i)^2}\right)/(n-1)$$

The final result is achieved if $(1/n)\Sigma_i(1/e_i)\mu$ is subtracted from (6.17) and completes the proof.

It is straightforward to construct an unbiased estimator of τ^2 by

$$\hat{\tau}^2_{corrected} = (\hat{\tau}^2 - \alpha\hat{\mu})/\beta \qquad (6.18)$$

where the non-random constants α and β are provided in (6.16).

To illustrate this estimation process, we consider the following example.

Example 6.8: Table 6.4 gives the observed and expected Hepatitis B cases in 23 city regions of Berlin. A test for heterogeneity of type (6.4) is given by $\chi^2 = \Sigma_i(O_i - \hat{\mu} \, e_i)^2/(\hat{\mu} \, e_i)$. Using the pooled estimate $\hat{\mu} = \Sigma_i O_i/\Sigma_i e_i = 1.019$, we get $\chi^2 = 193.52$ with 22 df, clearly indicating heterogeneity. Table 6.5 gives the two estimates of τ^2 using the arithmetic mean and the pooled mean estimate of μ. The estimated variances of the observed SMRs are also given. In both situations, we see that $\hat{\tau}^2$ corresponds to a high percentage of the estimated variance of observed SMRs, indicating the contribution of the heterogeneity distribution to the variance of the SMRs.

Area i	O_i	e_i	Area i	O_i	e_i
1	29	10.7121	13	15	8.3068
2	26	17.9929	14	11	15.6438
3	54	18.1699	15	11	11.8289
4	30	19.2110	16	2	9.9513
5	16	21.9611	17	2	10.8313
6	15	14.6268	18	9	18.3403
7	6	9.6220	19	2	5.1758
8	35	17.2671	20	3	10.9543
9	17	18.8230	21	11	20.0121
10	7	18.2705	22	5	13.8389
11	43	32.1823	23	2	12.7996
12	17	24.5929	–	–	–

Table 6.4. Observed and expected Hepatitis B cases in 23 City Regions of Berlin (Source: Berlin Census Bureau, 1995)

Estimator	$\hat{\mu}$	$\hat{V}ar$ (SMRs)	$\hat{\tau}^2$	Ratio of $\hat{\tau}^2$ to $\hat{V}ar$ (SMRs)
Simple mean	0.9751	0.6214	0.5489	0.883
Pooled mean	1.0188	0.6234	0.5470	0.877

Table 6.5. Estimates of the mean and variance of the SMRs and $\hat{\tau}^2$ for the SMRs for Hepatitis B in 23 city regions in Berlin

Proportion data. Suppose we have a sample of n proportions $r_1 = X_1/m_1$, ..., $r_n = X_n/m_n$, where X_i is the number of events, and m_i the number at risk. According to the binomial example we have

$$\text{Var}(X) = E(\sigma^2(\lambda)) + m^2\tau^2 = m\mu(1 - \mu) + m(m - 1)\tau^2.$$

Note that $\text{Var}(X) \approx m\mu + m^2\tau^2$, if μ is small and m is large, the variance based on the Poisson approximation of the binomial. Thus, one could use the estimate (6.15) with e_i replaced by m_i and O_i replaced by X_i as an estimate of τ^2 in this case. In case the approximation is *not* valid, one could use the formula

$$\hat{\tau}^2 = \frac{1}{n-1}\Sigma_i\frac{(X_i - m_i\hat{\mu})^2}{m_i(m_i - 1)} - \frac{1}{n}\hat{\mu}(1 - \hat{\mu})\Sigma_i\frac{1}{m_i - 1}. \qquad (6.19)$$

Similarly, we can use two estimators of μ, the simple mean and the pooled mean in (6.19). Equation (6.19) is simply an extension of the earlier formula we used for equal sample sizes $(m_i = m)$.

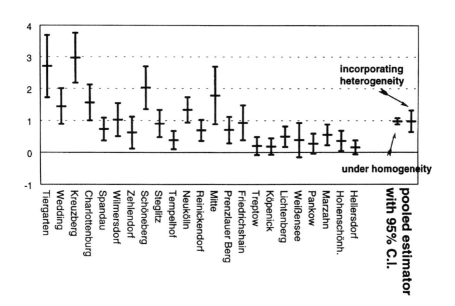

Figure 6.5. SMR estimates of Hepatitis B in 23 Berlin city areas with 95% pointwise confidence intervals.

Example 6.9: These formulas are illustrated at the SIDS-data of 100 North Carolina counties (Symons *et al.* 1982). Using the pooled estimator, the estimated mean rate is $\hat{\mu} = \Sigma_i X_i / \Sigma_i m_i = 2.02256$ per 1000 live births. We find $\chi^2 = \Sigma_i (X_i - \hat{\mu}\, m_i)^2 / (\hat{\mu}\, m_i) = 225.54$ (99 df), indicating heterogeneity. The results of our calculations for $\hat{\tau}^2$ are given in Table 6.6. We have in both situations a value of $\hat{\tau}^2$ which is more than one third of the total variance of the rates.

Estimator	$\hat{\mu}$	$\hat{\text{Var}}$ (Rates)	$\hat{\tau}^2$	Ratio of $\hat{\tau}^2$ to $\hat{\text{Var}}$ (Rates)
Simple mean	2.04631	2.47569	0.90803	0.367
Pooled mean	2.02256	2.47626	0.92677	0.374

Table 6.6. Estimates of the mean and variance of the rates and $\hat{\tau}^2$ for the rates of SIDS in 100 North Carolina counties (rates expressed in per 1000 population)

Example 6.10: To demonstrate these ideas we look at a data set on herd infection with trypanosomiasis discussed in Böhning and Greiner (1998). See also Chapter 8, Section 8.4. The data were collected in Mukono County which is located in the south-eastern part of Uganda and covers an area of approximately 200 km². The data (listed in Table 6.7) stem from a cross-sectional pilot study launched and accomplished in June/July 1994 for a project on trypanocide resistance in the peri-urban dairy production near Kampala. The sampling frame consisted of 187 dairy farms existing in the region (information from census April 1994) from which 50 farms were selected at random using random number tables, stratified for three categories of herd size: (small (1 – 10 cattle), medium (11–30), large (more than 30)). A total of 487 cattle was sampled on the identified farms. Using formula (6.19) we find $\hat{\tau}^2$ = 0.0039 which is about 14% of the observed variance $(1/(n - 1))\Sigma_i (X_i - m_i \hat{\mu})^2 / (m_i(m_i - 1))$ using $\hat{\mu} = (1/n)\Sigma_i\, X_i / m_i$.

As in the equal sample size case, $\hat{\tau}^2$ is *not* unbiased for τ^2 for *both* the simple mean and the pooled mean, although the bias disappears with either big n or ms.

Theorem 6.8: Let X_1, ..., X_n be a random sample of Binomial counts with associated sample sizes m_1, ..., m_n, e.g., $X_i \sim \text{Bi}(m_i, \lambda)$ where λ has distribution P with mean μ and variance τ^2. Then we have the following properties for the estimator (6.19):

(a) If $\hat{\mu} = \dfrac{1}{n}\Sigma_i \dfrac{X_i}{m_i}$, then

Farm	Cases	Sample size	Infection rate	Farm	Cases	Sample size	Infection rate
1	4	9	0.44	26	1	7	0.14
2	0	5	0	27	1	3	0.33
3	3	9	0.33	28	1	11	0.09
4	14	32	0.44	29	1	3	0.33
5	2	17	0.12	30	1	3	0.33
6	0	3	0	31	1	9	0.11
7	1	4	0.25	32	4	9	0.44
8	3	17	0.18	33	0	9	0
9	0	7	0	34	0	7	0
10	0	15	0	35	3	19	0.16
11	0	8	0	36	1	13	0.08
12	0	12	0	37	0	12	0
13	0	9	0	38	5	18	0.28
14	0	16	0	39	2	11	0.18
15	6	16	0.38	40	0	12	0
16	2	5	0.40	41	0	2	0
17	0	9	0	42	2	7	0.29
18	0	6	0	43	2	7	0.29
19	2	8	0.25	44	4	10	0.40
20	0	6	0	45	3	10	0.30
21	0	3	0	46	1	3	0.33
22	1	7	0.14	47	1	15	0.07
23	1	8	0.13	48	0	6	0
24	0	10	0	49	1	6	0.17
25	12	28	0.43	50	1	6	0.17

Table 6.7. Number of cattle infected with Trypanosoma spp., sample size and infection rate for 50 dairy farms in Mukono County, Uganda (data from June 1994, total sample size 487)

$$
\begin{aligned}
& E\left\{\left(\frac{1}{n-1}\Sigma_i\frac{(X_i-m_i\hat{\mu})^2}{m_i(m_i-1)}\right)-\frac{1}{n}\hat{\mu}(1-\hat{\mu})\Sigma_i\frac{1}{m_i-1}\right\} \\
& =\left(\frac{n-2}{n-1}+\frac{\Sigma_i\Sigma_j\frac{m_i(m_j-1)}{(m_i-1)m_j}}{n^2(n-1)}+\frac{\Sigma_i\Sigma_j\frac{m_j-1}{(m_i-1)m_j}}{n^3}\right)\tau^2 \qquad (6.20) \\
& +\left(\frac{\Sigma_i\Sigma_j\frac{m_i}{(m_i-1)m_j}}{n^2(n-1)}+\frac{\Sigma_i\Sigma_j\frac{1}{(m_i-1)m_j}}{n^3}-\frac{\Sigma_i\frac{1}{m_i-1}}{n(n-1)}\right)\mu(1-\mu).
\end{aligned}
$$

(b) If $\hat{\mu} = \dfrac{\Sigma_i X_i}{\Sigma_i m_i}$, then

$$E\left\{\left(\frac{1}{n-1}\Sigma_i\frac{(X_i-m_i\hat{\mu})^2}{m_i(m_i-1)}\right) - \frac{1}{n}\hat{\mu}(1-\hat{\mu})\Sigma_i\frac{1}{m_i-1}\right\}$$

$$= \left(\frac{n-2}{n-1} + \frac{\Sigma_i\Sigma_j\dfrac{m_im_j(m_j-1)}{m_i-1}}{(\Sigma_i m_i)^2} + \frac{\Sigma_i\Sigma_j\dfrac{m_j(m_j-1)}{m_i-1}}{n(\Sigma_i m_i)^2}\right)\tau^2 \qquad (6.21)$$

$$+ \left(\frac{\Sigma_i\dfrac{1}{m_i-1}}{n(n-1)} + \frac{\Sigma_i\dfrac{1}{m_i-1}}{n(\Sigma_i m_i)} - \frac{\Sigma_i\dfrac{m_i}{m_i-1}}{(n-1)(\Sigma_i m_i)}\right)\mu(1-\mu)$$

Proof.

(a) We write $(X_i - m_i\hat{\mu})^2$ as $(X_i - m_i\mu + m_i\mu - m_i\hat{\mu})^2$ and work out the expected values of the three terms:

$$E(X_i - m_i\mu)^2 = m_i\mu(1-\mu) + m_i(m_i-1)\tau^2$$

$$E\,2(X_i - m_i\mu)(m_i\mu - m_i\hat{\mu}) = -2/n[m_i\mu(1-\mu) + m_i(m_i-1)\tau^2]$$

$$E(m_i\mu - m_i\hat{\mu})^2 = \frac{1}{n^2}m_i^2\,[\Sigma_j 1/m_j\,\mu(1-\mu) + \Sigma_j(m_j-1)/m_j\,\tau^2]$$

Summing these terms up over i and dividing by $(n-1)$ leads to:

$$E\left(\frac{1}{n-1}\Sigma_i\frac{(X_i-m_i\hat{\mu})^2}{m_i(m_i-1)}\right) = \left[\frac{1}{n-1}\Sigma_i 1/(m_i-1) - \frac{2}{n(n-1)}\Sigma_i 1/(m_i-1)\right.$$

$$+ \left.\frac{1}{n^2(n-1)}\Sigma_i\frac{m_i}{m_i-1}\Sigma_j 1/m_j\right]\mu(1-\mu) + \left[n/(n-1) - 2/(n-1)\right.$$

$$+ \left.\frac{1}{n^2(n-1)}\Sigma_i\frac{m_i}{m_i-1}\Sigma_j\frac{m_j}{m_j-1}\right]\tau^2$$

From this expected value we have to subtract the expected value

$$E\left(\frac{1}{n}\hat{\mu}(1-\hat{\mu})\Sigma_i\frac{1}{m_i-1}\right) = \frac{1}{n}\Sigma_i\frac{1}{m_i-1}\left[\mu - \text{Var}\left(\frac{1}{n}\Sigma_i\frac{X_i}{m_i}\right) - \mu^2\right].$$

Since

$$Var\left(\frac{1}{n}\Sigma_i\frac{X_i}{m_i}\right) = \frac{1}{n^2}\Sigma_i\frac{1}{m_i}\mu(1-\mu) + \frac{1}{n^2}\Sigma_i\frac{m_i-1}{m_i}\tau^2 ,$$

we find that

$$E\left(\frac{1}{n}\hat\mu(1-\hat\mu)\Sigma_i\frac{1}{m_i-1}\right) = \left[\frac{1}{n}\Sigma_i\frac{1}{m_i-1} - \frac{1}{n^3}\Sigma_i\frac{1}{m_i-1}\Sigma_i\frac{1}{m_i}\right]\mu(1-\mu)$$
$$- \frac{1}{n^3}\Sigma_i\frac{1}{m_i-1}\Sigma_i\frac{m_i-1}{m_i}\tau^2 .$$

Now we can compute the difference between

$$E\left(\frac{1}{n-1}\Sigma_i\frac{(X_i-m_i\hat\mu)^2}{m_i(m_i-1)}\right)$$

and

$$E\left(\frac{1}{n}\hat\mu(1-\hat\mu)\Sigma_i\frac{1}{m_i-1}\right)$$

which leads to the result (6.20).

(b) can be worked out analogously.

The results (6.20) and (6.21) can be used to develop bias-corrected esti-mators of τ^2 by means of a correction of the form $[\hat\tau^2 - \alpha\hat\mu (1 - \hat\mu)]/\beta$ where β and α are the coefficients of τ^2 and $\mu(1 - \mu)$ in (6.20) and (6.21), respectively. Note, however, that this correction will *not* lead to a com-pletely bias-free estimator since $E\{\hat\mu (1 - \hat\mu)\} = \mu(1 - \mu) - Var(\hat\mu) \neq \mu(1 - \mu)$. Also, (6.20) and (6.21) are useful in determining the amount of possible bias that can occur in a given situation. Let $\hat\mu$ be the simple mean. Then, for example, for the data of Example 6.10 we find for the coefficient α of $\mu(1 - \mu)$: 0.0002, and for the coefficient β of τ^2: 1.0049. If the bias correction $[\hat\tau^2 - \alpha\hat\mu (1 - \hat\mu)]/\beta$ is used we find

$$E([\hat\tau^2 - \alpha\hat\mu(1-\hat\mu)]/\beta) = \tau^2 + \alpha/\beta Var(\hat\mu)$$

$$= \tau^2 + (\alpha/\beta)\frac{\Sigma_i\mu(1-\mu)/m_i + \Sigma_i(m_i-1)/m_i\tau^2}{n^2}$$

This gives improved bias-corrected coefficients for τ^2: $1 + \alpha/\beta(\Sigma_i(m_i - 1)/m_i)/n^2$ and for $\mu(1 - \mu)$: $+ \alpha/\beta(\Sigma_i 1/m_i)/n^2$. For the data of Example 6.10, they are $1 + 0.000000646$ and $+ 0.000003745$. Practically, the estimator is now bias-free.

In the case of equal binomial denominators $m_i = m$ for all $i = 1, ..., n$ we had found the bias-corrected estimator (6.11)

$$\hat{\tau}^2_{corrected} = \frac{nm - 1}{m^2(m-1)n}S^2 - \frac{\bar{X}}{m}\left(1 - \frac{\bar{X}}{m}\right)/(m-1)$$

An extension of this adjustment to the case of unequal binomial denominators leads to

$$\hat{\tau}^2_{corrected} = \frac{1}{n(n-1)}\Sigma_i(nm_i - 1)(X_i - m_i\hat{\mu})^2/[m_i^2(m_i - 1)] \quad (6.22)$$
$$- \frac{1}{n}\hat{\mu}(1 - \hat{\mu})\Sigma_i\frac{1}{m_i - 1}$$

When $\hat{\mu}$ is estimated by the simple mean, (0.22) has a smaller bias compared to $\hat{\tau}^2$ in (6.19). In Example 6.9, there is negligible difference between the estimators given in (6.19) and in (6.22) due to the large ms (minimum 284 and maximum 21,588) and n. The use of the Poisson approximation also results in very similar values.

6.5   Consequences for efficient estimation of the mean μ of the mixing distribution

Typically, for computing a confidence interval for μ the variance of $\hat{\mu}$ is needed. We first determine for SMR data ($X \sim Po(\lambda e)$) the variances of both the simple mean and the pooled mean. We find that

$$\text{Var}\left(\frac{1}{n}\Sigma_i\frac{O_i}{e_i}\right) = \mu\left(\frac{1}{n^2}\Sigma_i\frac{1}{e_i}\right) + \tau^2\left(\frac{1}{n}\right) \quad (6.23)$$

$$\text{Var}\left(\frac{\Sigma_i O_i}{\Sigma_i e_i}\right) = \mu\left(\frac{1}{\Sigma_i e_i}\right) + \tau^2\left(\frac{\Sigma_i e_i^2}{(\Sigma_i e_i)^2}\right) \quad (6.24)$$

There are several things noteworthy here. In each of the variance expressions, the first term is the usual variance formula for $\hat{\mu}$ under the assumption of *homogeneity* ($\tau^2 = 0$). From these expressions, we see the contribution of the *heterogeneity* distribution to the variance of $\hat{\mu}$. (Incidentally, if we replace the parameters with their estimates in the case of homogeneity, (6.24) leads to the usual textbook formula for the variance of the simple mean of SMRs. See for example Lilienfeld and Lilienfeld (1980, p. 353).)

We also note that a comparison of the variances of these two estimators is accomplished by looking at the coefficients of μ and τ^2. It is shown in Theorem 6.8 that the coefficient of μ for the pooled mean is smaller than that of the sample mean. The reverse is true for the coefficient of τ^2. Thus, if there is *no* heterogeneity, that is $\tau^2 = 0$, we see that the variance of the pooled mean is smaller than that of the simple mean. But if τ^2 is becoming large enough and is dominating the variance of $\hat{\mu}$, then the simple mean has a smaller variance compared to the pooled mean.

Theorem 6.9: For non-negative numbers $e_1, \ldots, e_n$ we have:

(a)* $\dfrac{1}{n^2}\Sigma_i\dfrac{1}{e_i} \geq \dfrac{1}{\Sigma_i e_i}$

(b) $\dfrac{\Sigma_i e_i^2}{(\Sigma_i e_i)^2} \geq \dfrac{1}{n}$

Proof.

(a) We note that for any *convex* function $f(x)$ the following inequality holds:

$$\Sigma_i\, \alpha_i\, f(x_i) \geq f\,(\Sigma_i\, \alpha_i\, x_i),$$

where $\alpha_i \geq 0$ and $\Sigma_i\, \alpha_i = 1$. Now, let $f(x) = 1/x$, and it follows that

$$\frac{1}{n}\Sigma_i\frac{1}{e_i} \geq \frac{1}{\frac{1}{n}\Sigma_i e_i} = \frac{n}{\Sigma_i e_i},$$

with $x_i = e_i$ and $\alpha_i = 1/n$, or

---

* A corollary to this theorem is that the arithmetic mean is at least as large as the harmonic mean.

$$\frac{1}{n^2}\Sigma_i\frac{1}{e_i} \geq \frac{1}{\Sigma_i e_i}\,.$$

(b) $\Sigma_i e_i{}^2 - \dfrac{(\Sigma_i e_i)^2}{n} = \Sigma_i(e_i - \bar{e})^2 \geq 0.$

We do the same analysis for the proportion data and find similar conclusions. The variances of the estimators of μ are as follows:

$$\mathrm{Var}\!\left(\frac{1}{n}\Sigma_i\frac{X_i}{m_i}\right) = \mu(1-\mu)\!\left(\frac{1}{n^2}\Sigma_i\frac{1}{m_i}\right) + \tau^2\!\left(\frac{1}{n^2}\Sigma_i\frac{m_i-1}{m_i}\right)$$

$$\mathrm{Var}\!\left(\frac{\Sigma_i X_i}{\Sigma_i m_i}\right) = \mu(1-\mu)\!\left(\frac{1}{\Sigma_i m_i}\right) + \tau^2\!\left(\frac{\Sigma_i m_i(m_i-1)}{(\Sigma_i m_i)^2}\right)$$

These results lead us to the problem of determining optimal values of estimates for μ in the presence of heterogeneity. That is, using SMR data, we would like to find the optimal values of w_i such that $\mathrm{Var}\ (\Sigma_i w_i\ (O_i/e_i))$ is minimized. The weights that minimize this variance are those that are inversely proportional to the variances (see Chapter 5). The variance of (O_i/e_i) is given by

$$\mathrm{Var}\!\left(\frac{O_i}{e_i}\right) = \frac{e_i\mu + e_i^2\tau^2}{e_i^2} = \frac{\mu}{e_i} + \tau^2\,. \tag{6.25}$$

We first consider two extreme cases: when there is homogeneity and when the variance is totally dominated by τ^2. When homogeneity is assumed, the optimal values of w_i are the normed recripocals of the e_is, which leads to our pooled estimator $\hat{\mu} = \Sigma_i O_i/\Sigma_i e_i$. On the other extreme, when τ^2 totally dominates (μ/e_i) such that $\mathrm{Var}(O_i/e_i) \approx \tau^2$, then the optimal weights tend to go to $1/n$. Thus, the simple mean is an optimal solution in this situation. When *neither* of these two situations is true, then we consider weights derived from (6.25). However, we still need to estimate the μ and $\tau^{2}.$* Here, we propose an iterative solution, an *iterative reweighted least squares algorithm*, for finding the optimal weights and, eventually, for finding the optimal values of μ and τ^2.

Algorithm 6.1 (for SMR data):

Step 0. Choose some initial value of $\mu = \mu_0$ ($= \overline{\mathrm{SMR}}$). Based on this value, compute $\hat{\tau}^2(\mu) = \hat{\tau}^2(\mu_0) = \tau^2{}_0$ as given in (6.14) for SMR data.

---

* Actually, it is one of the remarkable properties of the two extreme situations that the optimal weights become *independent* of μ and τ^2.

Step 1. Compute the weights $w_i = [1/(\mu/e_i + \hat{\tau}^2(\mu))]$.

Step 2. Determine $\mu = \Sigma_i w_i(O_i/e_i)/\Sigma_i w_i$ and $\tau^2 = \hat{\tau}^2(\mu)$.

Step 3. Go to step 1.

Step 4. Repeat this process until convergence is attained.

For proportion data the algorithm 6.1 needs to be changed only for computing the weights $w_i = 1/[\mu(1 - \mu)/m_i + (m_i - 1)/m_i \ \hat{\tau}^2(\mu)]$. For the heterogeneity variance, we can use the non-adjusted version of (6.19) for computing $\hat{\tau}^2(\mu)$:

$$\hat{\tau}^2 = \frac{1}{n}\Sigma_i \frac{(Y_i - m_i\mu)^2}{m_i(m_i - 1)} - \frac{1}{n}\mu(1 - \mu)\Sigma_i \frac{1}{m_i - 1} \qquad (6.26)$$

We illustrate this with one example from meta-analysis.

Example 6.11: *(Rate of Agoraphobia)* Frequently, there is need to combine results of various studies on a certain research question of interest, such as the worldwide distribution on agoraphobia as studied by Eaton (1995). Table 6.8 gives the result on $n = 7$ studies on the prevalence of agoraphobia (the fear and anxiety connected with open, public places connected with the appearance of masses of people).

Study i	AP X_i	Sample size X_m	Prevalence rate X_i/X_m ($\times$ 1000)
1	808	14436	55.9712
2	78	1366	57.1010
3	107	1551	68.9877
4	94	3258	28.8521
5	66	3134	21.0593
6	71	1966	36.1139
7	429	8098	52.9760

Table 6.8. Prevalence rates of agoraphobia based on 7 studies as reported in Eaton (1995)

Using the pooled estimate, $\hat{\mu} = 48.8923$ per 1000 persons. Here, we find $\chi^2 = \Sigma_i(x_i - \hat{\mu} m_i)^2/(\hat{\mu} m_i) = 115.24$ (6 df), indicating large heterogeneity. The comparison of the variances of the simple arithmetic mean, pooled mean, and the iterated mean of the rates is given in Table 6.9. The pooled mean has an efficiency of only 54.1% while the simple mean is

Estimator	Estimate	Var (estimate)	Efficiency	$\hat{\tau}^2$	Ratio of $\hat{\tau}^2$ to $\hat{\text{Var}}$ (SMRs)
simple	45.8659	43.3111	99.3%	285.4582	0.944
pooled	48.8923	79.5503	54.1%	295.0823	0.976
optimal	45.7361	43.0008	1	284.5935	0.942

Table 6.9. Comparison of variances of the simple, pooled, and iterated means of the rate of agoraphobia (rates expressed in per 1000 population)

almost as efficient as the optimal mean. Here is a case where the variance of the estimate is almost solely attributable to the heterogeneity distribution. Figure 6.6 shows how much the variance can be underestimated by ignoring heterogeneity if it exists.

Meta-Analysis of Agoraphobia

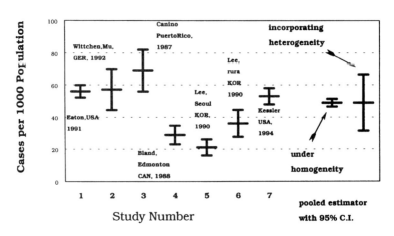

Figure 6.6. Confidence intervals for the rate of agoraphobia in 7 studies and from the results of meta-analysis assuming homogeneity and incorporating heterogeneity (Eaton, 1995).

It might be of interest to know more about the iterative reweighted least squares algorithm 6.1. In fact, a proof of convergence is still missing. In practice, the algorithm converges quickly, independent of initial values. Table 6.10 illustrates the convergence process for 3 different initial values of μ. The algorithm was stopped when 4 significant digits of accuracy were reached.

	Initial value	Iteration 1	Iteration 2	Iteration 3
μ	0.5	0.9777	0.9817	0.9817
τ^2	–	0.7830	0.5217	0.5214
μ	0	0.9751	0.9817	0.9817
τ^2	–	1.5453	0.5219	0.5214
μ	10	0.9756	0.9817	0.9817
τ^2	–	81.2992	0.5218	0.5214

Table 6.10. Convergence behavior of algorithm 6.1

6.6 Comparison to the nonparametric likelihood approach

The question arises of how the current approach compares to other nonparametric approaches. One alternative possibility consists of finding the nonparametric maximum likelihood estimator $\hat{P}$ (NPMLE) in the marginal distribution $f(x, P) = \int_{-\infty}^{\infty} f(x, \lambda) P(d\lambda)$ which has been studied in detail in Chapters 2 and 3. It was shown that this NPMLE $\hat{P}$ is always discrete having finite mass points $\lambda_1, ..., \lambda_k$ and weights $p_1, ..., p_k$, and can be computed with existing software such as C.A.MAN (Böhning et al. 1992). After having done so the variance of $\hat{P}$ is provided by

$$\mathrm{Var}(\hat{P}) = \Sigma_j p_j (\lambda_j - \bar{\lambda})^2, \bar{\lambda} = \Sigma_j p_j \lambda_j. \qquad (6.27)$$

This has been done for the hepatitis data studied in Example 6.8. The results are provided in Table 6.11. The variance according to (6.27) is 0.5332, which compares well with the estimates of τ^2 given in Table 6.5.

Parameter	Values				
Means	λ_1	λ_2	λ_3	λ_4	λ_5
	0.3111	0.6747	1.4347	1.8751	2.7760
Weights	p_1	p_2	p_3	p_4	p_5
	0.2574	0.4066	0.1973	0.0450	0.0938
Variance of $\hat{P}$	0.53320				

Table 6.11. Maximum likelihood estimate of P with associated variance for Hepatitis cases in 23 City Regions of Berlin in 1995

In a similar way, one can estimate the heterogeneity distribution for the SIDS data of Example 6.9. The results are provided in Table 6.12. Note here that the heterogeneity consists of four sub-populations receiving different weights. The estimated variance according to (6.27) is 1.1552×10^{-6} which again compares well to the estimates given in Example 6.9 in Table 6.6.

Parameter	Values			
Means	λ_1	λ_2	λ_3	λ_4
	0.001250	0.002074	0.003739	0.009008
Weights	p_1	p_2	p_3	p_4
	0.3215	0.5156	0.1521	0.0108
Variance of $\hat{P}$	1.1552×10^{-6}			

Table 6.12. Maximum likelihood estimate of P with associated variance for SIDS data of North Carolina

Finally, one can estimate the heterogeneity distribution for the agoraphobia data of Example 6.11. The results are provided in Table 6.13. Note here that the heterogeneity consists of five subpopulations receiving different weights, whereas the sample size of 7 corresponds to the number of studies. This expresses the strength of population heterogeneity already expressed in Table 6.9. The estimated variance according to (6.27) is 244.848, which corresponds to 81.0% of the variance of the rates, whereas the proposed moment estimators range around 95%. It remains to be investigated if in the case of strong heterogeneity and small sample size the variance of the nonparametric maximum likelihood estimator of the heterogeneity distribution tends to underestimate the heterogeneity variance.

Parameter	Values				
Means	λ_1	λ_2	λ_3	λ_4	λ_5
	21.154	30.726	34.553	55.201	68.168
Weights	p_1	p_2	p_3	p_4	p_5
	0.1437	0.2088	0.0759	0.4494	0.1222
Variance of $\hat{P}$	244.848				

Table 6.13. Maximum likelihood estimate of P with associated variance for meta-analysis of agoraphobia in 7 studies

In addition to the nonparametric approach, there is the parametric model: for SMR-data the heterogeneity distribution is frequently modeled by a gamma distribution, whereas for proportion-data the beta distribution is often used for the heterogeneity distribution. After estimating their two parameters one can easily compute their variances. Often these models are used for reasons of their mathematical convenience, not necessarily because they are the right ones. Clearly the nonparametric approach offers more flexibility in that it imposes less structure on the heterogeneity distribution.

CHAPTER 7

C.A.MAN-application: disease mapping

7.1 Conventional approach I: mapping percentiles

Environmental justice and equity are emerging concepts in the development of environmental health policy. These concepts are related to questions on the spatial distribution of environmental contaminants in the population leading to the potential occurrence of certain diseases in different parts of the population. Disease mapping can be defined as a method for displaying the spatial distribution of disease occurrence (or exposure occurrence), the most prominent forms being the variety of existing cancer atlases (Holland 1991, 1992; Becker *et al.* 1984; Becker and Wahrendorf 1997; Cartwright *et al.* 1990). Typically, the measure of interest such as the *standardized mortality ratio* SMR* has been computed only for some aggregated unit such as an area (county, municipality, etc.). Let $x_1, ..., x_n$ be the sample for the n areas under consideration. Then the classification of each area is based on the percentiles of the empirical distribution of the $x_1, ..., x_n$. Recall that the *pth percentile* of some continuous random variable X with distribution function F is given as that value x_p in the sample space of X, for which $F(x_p) = p$, $p \in (0, 1)$. For $p = 1/2$ we have the *median*; for $p = 1/4, 1/2, 3/4$ the *quartiles*; for $p = 0.2, 0.4, 0.6, 0.8$ the *quintiles*; and so forth. In disease map construction often the quintiles or septiles are used. In Figure 7.1 we see a European map on stomach cancer based on quintiles. This means that each of the areas is classified according to its SMR-value into the associated quintile.

---

* The SMR is defined as the ratio of observed O and expected e mortality cases, where the number of expected cases is computed on the basis of an external reference population. In more detail: let n_l denote the number at risk in the lth age group of the study population and μ_l the mortality rate in the lth age group of the *reference population*, then $e = n_1\mu_1 + ... + n_k\mu_k$, assuming that there are k age groups.

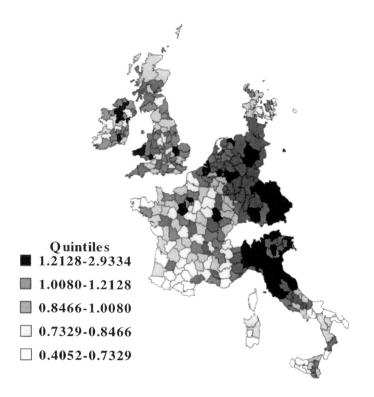

Quintiles
- ■ 1.2128-2.9334
- ▨ 1.0080-1.2128
- ▧ 0.8466-1.0080
- □ 0.7329-0.8466
- □ 0.4052-0.7329

Figure 7.1. Spatial distribution of stomach cancer for males in Europe using a classification of the SMR-distribution based upon quintiles.

As can be seen in Figure 7.1 there is a concentration of higher relative risk as measured on the SMR-scale in Southwest Germany (Bavaria) and Northern Italy, whereas there is lower relative risk in other parts of Germany and Europe. The idea behind disease mapping is often to represent that part of spatial variation which cannot be explained by *known* factors, nor is due to *random variation*. By doing so, the epidemiologist hopes to find hints for *unknown* risk factors. A common method to screen for risk factors is to compare the disease map, say stomach cancer, on the one hand with an exposure map, say a nutritional habit, on the other hand. This analysis points to a direction of epidemiological investigation, called *ecological studies*, in which aggregated data (in contrast to individual patient or person data) are related to each other. Often this level of aggregation is a geographical unit or area. The interested reader is referred to the work of Lawson *et al.* (1998), Neutra (1998), or the review papers by Best (1998) and Biggeri *et al.* (1998). In any case, if regions with high risk are detected in connection with suspected risk factors, further epidemiological investigations are required to reach further conclusions. Returning to the example of stomach cancer mentioned above, Boeing and Frentzel-Beyme (1991) as well as Boeing *et al.* (1991) undertook case-control studies to investigate potential risk factors in further detail. Cases and controls were sampled from Bavaria (high risk) and Hessia (low risk) to have the most variation in disease occurrence. The following risk factors were studied: *type of drinking water, consumption of smoked meat (meat was smoked at home), usage of refrigerator,* and *vitamin-C consumption.* The relative risks in Table 7.1 were established by means of odds ratios.

As can be seen from Table 7.1, vitamin-C consumption is preventive for stomach cancer whereas the use of home-fountain water or eating meat which was smoked at home leads to an increased relative risk. All these factors relate to the pathophysiological hypothesis that they lead to an increased occurrence of nitrosamines which are known

Risk factor	Odds ratio	95% C.I.
Drinking water		
Own fountain	2.18	1.39–3.42
Central	1	
Smoking meat at home		
yes	3.33	1.56–7.12
no	1	
Refrigerator–years		
< 25 years	1.33	0.82–2.15
25–30 years	1.12	0.70–1.80
> 30 years	1	
Vitamin–C consumption		
1. quintile	2.32	1.22–4.43
2. quintile	1.69	0.88–3.26
3. quintile	1.60	0.82–3.10
4. quintile	1.40	0.71–2.73
5. quintile	1	

Table 7.1. Odds ratios with 95% C.I. for various risk factors

to be carcinogenic. For a detailed discussion on the matter see Boeing and Frentzel-Beyme (1991) and Boeing *et al.* (1991).

It is quite clear from the construction of the map that every choice of percentile will *force* a certain pattern or structure in the map. We see in Figure 7.2 a disease map for the same SMR-distribution, but in this case the median was used for classification. It is quite clear from these maps that the percentile method has its deficiencies.

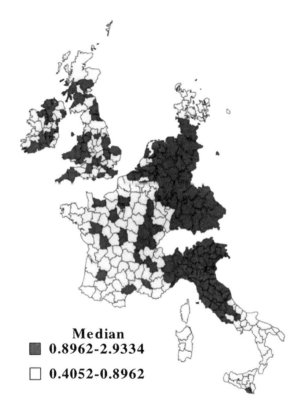

Figure 7.2. Spatial distribution of stomach cancer for males in Europe using a classification of the SMR-distribution based upon the median (above or below median SMR = 0.8962).

7.2   Conventional approach II: mapping *P*-values

Another conventional method used frequently to construct disease atlases (for example, see *Leukemia and Lymphoma: An atlas of distribution within areas of England and Wales 1984–1988*, Cartwright *et al.* 1990) is also based on the SMR = O/e, where O is the observed number of death cases and e is the expected number of death cases for a given region. However, it is now assumed that in area i the observed number of death cases O_i follows a Poisson with parameter λe_i:

$$f(x_i, \lambda) = Po(o_i, \lambda e_i) = \exp(-\lambda e_i)(\lambda e_i)^{o_i}/o_i! \qquad (7.1).$$

Here, $x_i = o_i/e_i$ is the *observed SMR* in area i, $i = 1, ..., n$, whereas λ is the *theoretical SMR*. The conventional display map is based on a classification using the *P-value* under the *homogeneous* Poisson distribution:

$$P(O_i \geq o_i) = Po(o_i, \lambda e_i) + Po(o_i + 1, \lambda e_i) + ..., \text{ if } x_i \geq \lambda$$
$$P(O_i < o_i) = Po(o_i - 1, \lambda e_i) + Po(o_i - 2, \lambda e_i) +$$
$$... + Po(0, \lambda e_i), \text{ if } x_i < \lambda.$$

λ is either set to 1 (no increased risk) or replaced by the MLE under homogeneity $\hat{\lambda} = \sum_{i=1}^{n} O_i / \sum_{i=1}^{n} e_i$. In recent years, a WINDOWS program by the name DISMAP has been developed out of C.A.MAN (Schlattmann and Böhning 1993; Schlattmann, Dietz, and Böhning 1996), *solely for the purpose of disease mapping.* Figure 7.3 shows a display of stomach cancer, based on the classification using *P*-values ($\lambda = 1$). A map with another pattern arises. Also, this method has a number of deficiencies. For one, the significance of some SMR will depend much on the size of the area, in the sense that areas with small population size have greater chances to show a significant result. This can be seen in Figure 7.3, where a number of northern areas are now in the darker colors and the very articulate image of Figure 7.1 is lost. Second, the method is *unable* to detect a homogeneous population, as is shown in the following example.

Example 7.1: Consider a population with homogeneous relative risk λ in *all* areas and consider the classification rule in which an area is classified into group *normal* if its *P*-value is $\geq \alpha$, otherwise into the group *not normal*. Evidently, Pr{ classification is correct } = $1 - \alpha$. If this is done for n areas, Pr{ classification is correct } = $(1 - \alpha)^n$, assuming independence. Replacing α with the Bonferroni-adjustment α/n will guarantee that

$$\Pr\{\text{classification is correct}\} = (1 - \alpha/n)^n \geq 1 - \alpha.$$

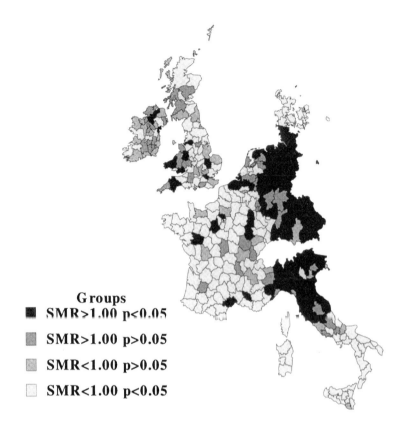

Groups
- SMR>1.00 p<0.05
- SMR>1.00 p>0.05
- SMR<1.00 p>0.05
- SMR<1.00 p<0.05

Figure 7.3. Disease map of stomach cancer according to a classification using P-values based on the homogeneous Poisson distribution.

In numbers: $n = 23$, $\alpha = 0.05$, $\alpha/n = 0.0022$, $(1 - \alpha/n)^n = 0.9512$. The question is what will happen to $(1 - \alpha/n)^n$ if n is becoming large. A Taylor series argument for the concave logarithm shows that

$$\log (1 + h) \leq \log(1) + h = h$$

for some h. Application to $h = -\alpha/n$ leads to the result

$$n \log(1 - \alpha/n) \leq n(-\alpha/n) = -\alpha$$

or, $$(1 - \alpha/n)^n \leq \exp(-\alpha) < 1,$$

which demonstrates that $(1 - \alpha/n)^n$ is bounded away from 1. As a conclusion it can be stated that the probability $(1 - \alpha/n)^n$ of a correct classification will remain different from 1, even if n is approaching infinity. Therefore, this hypothesis testing-based estimation procedure of disease map structure is *inconsistent* (Schlattmann and Böhning 1993).

Clayton and Kaldor (1987) formulate the critical issues: *One of the main problems has been the choice of the appropriate measure of cancer incidence or mortality to map. Some atlases have presented measures of relative risk, usually standardized mortality ratios (SMRs), while others display the statistical significance of local deviations of risk from the overall rates on the map.*

Both these approaches can badly misrepresent the geographical distribution of cancer incidence. In the former case, no account is taken of varying population size over the map, so that imprecisely estimated SMRs, based on only a few cases, may be the extremes of the map, and hence dominate its pattern. On the other hand, mapping significance alone totally ignores the size of the corresponding effect, so that on the map, two areas with identical SMRs may be indicated quite differently if they are of unequal population size, and the most extreme areas may simply be those with the largest populations. Not only are these approaches unsatisfactory, but the lack of a common format of presentation frustrates the comparison of cancer across atlases.

Obviously, sound statistical methodology for assessing environmental justice is required. On the one hand, we see as part of an information oriented society a strong development within so-called *geographic information systems* (GIS), usually without any built-in statistical methodology. On the other hand, there exists a diversity of spatial statistical methodology, where even the inexperienced statistical expert is easily lost. There exists the empirical Bayes approach with the prominent papers by Clayton and Kaldor (1989) and Manton *et al.* (1989). See also Louis (1984, 1991), Bernardinelli and Montomoli (1992), Biggeri *et al.* (1993), Cislaghi *et al.* (1995), and Tsutakawa, Shoop, and Marienfeld (1985). Marshall (1991) and Stigler (1990) point out the connection to shrinkage estimation à la Stein (Efron and Morris 1973). Clayton and Bernardinelli (1992) review Bayesian methods for modelling geographic disease rates. See also Besag and Mollié (1989). Devine and Louis (1994) consider *constrained* empirical Bayes approaches to protect against overshrinkage of the crude disease rates toward their grand mean. Lawson, Biggeri, and Dreassi (1998) discuss edge effects in disease mapping. Besag, York, and Mollié (1991) extend the Clayton and Kaldor (1987) approach by separating spatial from heterogeneity effects. Recently, there is an emerging desire to incorporate temporal effects into spatial modeling (Bernardinelli *et al.* 1995, Carlin and Louis 1996, Waller *et al.* 1997). It can be expected that there will be further development in this area and we are only able to touch upon a few issues here. In these fast developments of spatial methods and tools with increasing mathematical complexity, it is not surprising that the original public health user feels him/herself sometimes only marginally represented and, consequently, the public health user is turning to systems in which he/she can find simple answers, potentially erroneous ones. It will be a challenge of the future to provide appropriate and sound methods of disease mapping which are also simple and easy to understand even by the non-expert. In the present situation we are confronted with a number of disease maps which are often constructed by quite different methods (Walter and Birnie 1991).

It was just this latter point of missing comparability of disease maps due to different methods that has led a group of people from various European countries,* headed by Dr. Andrew Lawson, to a collaborative research initiative to investigate and compare various disease mapping methods. These activities culminated in September 1997 in Rome in a joint workshop with WHO. As a result of this workshop a book (Lawson *et al.* 1998) was published, containing a collection of methods also used in disease mapping.** In the following we will highlight a few of these methods, though this collection is not meant to be representative.

7.3   Empirical Bayes estimators

It is one of the disadvantages of the two conventional methods presented in Sections 7.1 and 7.2 that they assume a *homogeneous* population, one single λ to be valid. It might be more *realistic* to assume that there is a distribution of λ valid in the population. This idea leads in a natural way to *empirical Bayes* estimates. We consider the *Bayes risk*

---

* The following countries were represented: Italy by Dr. Annibali Biggeri, Belgium by Dr. Emmanuel Lesaffre, France by Dr. Jean-Francois Viel, Germany by Dr. Dankmar Böhning, and Scotland by Dr. Andrew Lawson (project coordinator). The project was funded within the European community program BIOMED2 and was entitled Disease Mapping and Risk Assessment.

** A topic of its own right (though connected to disease mapping) is health risk assessment around putative sources (Hills and Alexander 1989; Lawson, Biggeri, and Williams 1998). For a case study on the connection of disease mapping and suspected point sources, see Hoffmann and Schlattmann (1998).

$$\iint\limits_{\lambda\ x}(\hat{\lambda}(x)-\lambda)^2 f(x|\lambda)\mathrm{d}x\ p(\lambda)\mathrm{d}\lambda \qquad (7.2)$$

with respect to the Euclidean loss function $(\hat{\lambda}-\lambda)^2$. Here $p(\lambda)$ denotes the distribution of λ in the population, which might be continuous or discrete. We recall that μ denotes the mean with respect to $p(\lambda)$ and τ^2 its variance. If x is a *standardized mortality ratio* $x = o/e$, for which O *conditional on* λ is Poisson with mean λe, then $E(X) = \lambda$ and $\mathrm{Var}(X) = \lambda/e$.

We are interested in finding a Bayes estimate $\hat{\lambda}(x)$ which minimizes (7.2). This can easily be accomplished by considering *linear Bayes estimators* $\hat{\lambda}(x) = \alpha + \beta\ x$.

Theorem 7.1: Given the situation above, the best linear Bayes estimator $\hat{\lambda}(x) = \alpha + \beta\ x$ is given by

$$\beta = \frac{\tau^2}{\tau^2 + \mu/e}\ \text{and}\ \alpha = (1-\beta)\mu. \qquad (7.3)$$

Proof. We work out expression (7.2) as

$$\iint\limits_{\lambda\ x}(\alpha+\beta x-\lambda)^2 f(x|\lambda)\mathrm{d}x\ p(\lambda)\mathrm{d}\lambda = \int\limits_{\lambda}[\beta^2\mathrm{Var}(x)+(\alpha+(\beta-1)\lambda)^2]p(\lambda)\mathrm{d}\lambda$$
$$= \int\limits_{\lambda}[\beta^2\lambda/e+(\alpha+(\beta-1)\lambda)^2]p(\lambda)\mathrm{d}\lambda$$
$$= \beta^2\mu/e+\alpha^2+2\alpha(\beta-1)\mu+(\beta-1)^2(\tau^2+\mu^2),$$

where we have used in the last expression that $\int_\lambda \lambda^2 p(\lambda)\mathrm{d}\lambda - \mu^2 = \tau^2$. This is a quadratic form in α and β, $Q(\alpha, \beta)$ say, which can be solved directly by taking partial derivatives and equating them to zero:

$$\partial Q/\partial\alpha = 2\alpha+2(\beta-1)\mu = 0$$
$$\partial Q/\partial\beta = 2\beta\mu/e+2\alpha\mu+2(\tau^2+\mu^2)(\beta-1) = 0$$

from which the result follows. This ends the proof.

The result has been derived before by various authors (Maritz and Lwin 1989; Carlin and Louis 1996; Marshall 1991). Note that it can be written as a convex combination of the prior mean μ and the individual SMR x: $\hat{\lambda}(x) = \alpha + \beta\ x = (1-\beta)\mu + \beta\ x$ with $\beta = \tau^2/(\tau^2 + \mu/e)$. Evidently, $\hat{\lambda}(x)$ will be close to x, if β is near 1, or if τ^2 is large. It will be near μ if β is close to 0, or if τ^2 is small. In this case, the Bayes estimate will show less variation in comparison to x. The x_i are *shrunk* to the prior mean.

This property puts these estimates in the *class of shrinkage-estimators*. For an introduction to shrinkage-estimators we refer the interested reader to Stigler (1990) who provides a Galtonian perspective, but also points out the connection to empirical Bayes estimation.

Example 7.2: We consider again the small area data of Martuzzi and Hills (1995) on perinatal mortality in the North West Thames Health Region in England based on the 5-year period 1986–1990. The epidemiological measure of interest is the standardized mortality ratio SMR. We use the following values for μ and τ^2: μ = 0.96 and τ^2 = 0.034 (the question how these were derived will be discussed further below). As can be seen in Figure 7.4 there is quite a large amount of shrinkage in the empirical Bayes estimate of the standardized mortality ratio, caused by the small value of the heterogeneity variance τ^2.

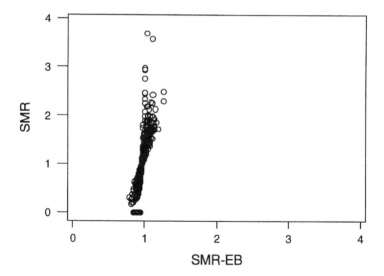

Figure 7.4. Scatterplot of x vs. $(1 - \beta)\mu + \beta x$.

It remains to discuss the important point of how estimates of μ and τ^2 can be achieved. In fact, it is this aspect which leads us to *empirical Bayes estimates*. Conventionally, the marginal density

$$f(o_i, e_i, \Phi) = \int_0^\infty \mathrm{Po}(o_i, \lambda e_i) p(\lambda) d\lambda$$

is considered and maximum likelihood estimates are found with respect to this mixture density. Here, *two* ways are possible. For one, $p(\lambda)$ could be a parametric, continuous density in which case Φ denotes its associated parameters. Second, one could leave $p(\lambda)$ completely nonparametric, in which case Φ would coincide with the nonparametric mixing distribution.

It was outlined in Chapter 1 (Example 1.3) that if $p(\lambda)$ takes specific forms, then specific marginal distributions are also achieved. For example, if $p(\lambda;\theta,\kappa) = \theta^{-\kappa}\lambda^{\kappa-1}e^{-\lambda/\theta}/\Gamma(\kappa)$ with $\mu = \kappa\theta$ and $\tau^2 = \kappa\theta^2 = \mu\theta = \mu^2/\kappa$ it follows that

$$f(o_i, e_i, \Phi) = \int_0^\infty \mathrm{Po}(o_i, \lambda e_i) p(\lambda) d\lambda$$

is the *negative binomial* distribution with mean μe_i and variance $\mu e_i + \tau^2 e_i^2$. Therefore, estimating μ and τ^2 leads to maximum likelihood estimation of the parameters of a negative binomial distribution. Unfortunately, there are no closed form solutions for this task and iterative procedures such as Newton–Raphson have to be used. See also Clayton and Kaldor (1987) for this point. Note that in this case the *empirical Bayes estimate* can be written also as $\mu + \tau^2/(\tau^2 + \mu/e)(x - \mu) = \kappa\theta + (\kappa\theta^2)/(\kappa\theta^2 + \kappa\theta/e)(x - \kappa\theta) = (o + \kappa)/(e + 1/\theta)$ (Marshall 1991). A simple *moment estimate* has been given in Chapter 6 for τ^2 in expression (6.15):

$$\hat{\tau^2} = \frac{1}{n-1}\Sigma_i(O_i - e_i\hat{\mu})^2/e_i^2 - \frac{1}{n}\hat{\mu}\Sigma_i\frac{1}{e_i} \qquad (7.4)$$

$$= \frac{1}{n-1}\Sigma_i(\mathrm{SMR}_i - \hat{\mu})^2 - \frac{1}{n}\hat{\mu}\Sigma_i\frac{1}{e_i}$$

with $\hat{\mu} = \frac{1}{n}\Sigma_i\frac{O_i}{e_i}$.

Example 7.3: We continue the discussion of Example 7.3. Using the estimate (7.4) we find for the North West Thames Health region data $\hat{\tau^2} = 0.034$ and $\hat{\mu} = 0.96$. This was used in Example 7.2 to compute the associated empirical Bayes estimates.

7.4   Mapping empirical Bayes estimates

In Section 1.4 of Chapter 1, the posterior distribution of λ had been defined as

$$f(\lambda|x) = \frac{f(x|\lambda)p(\lambda)}{f(x;P)} = \frac{f(x|\lambda)p(\lambda)}{\displaystyle\int_{-\infty}^{+\infty} f(x|\lambda)p(\lambda)d\lambda}$$

if P is a parametric distribution with density $p(\lambda)$, or

$$f(\lambda_j|x) = \frac{f(x|\lambda_j)p_j}{f(x;P)} = \frac{f(x|\lambda_j)p_j}{\displaystyle\sum_{l=1}^{k} f(x|\lambda_l)p_l}$$

if P is nonparametric and discrete. Having $f(\lambda\,|\,x)$ or its estimate available the posterior mean, which we called *the empirical Bayes* estimate of λ, is given as:

$$x^{\mathrm{EB}} = E(\lambda|x) = \int_{-\infty}^{+\infty} \lambda f(\lambda|x)d\lambda \qquad (7.5)$$

For the Poisson–Gamma model, (7.5) coincides with (7.3) as the following theorem shows.

Theorem 7.2: Let $f(x;\lambda) = Po(o, \lambda e) = \exp(-\lambda e)\,(\lambda e)^o/o!$ and $p(\lambda) = p(\lambda;\theta, \kappa) = \theta^{-\kappa}\lambda^{\kappa-1}e^{-\lambda/\theta}/\Gamma(\kappa)$. (Note that $\Phi = (\theta, \kappa)^{\mathrm{T}}$ and $\mu = \kappa\theta$, $\tau^2 = \kappa\theta^2 = \mu\theta = \mu^2/\kappa$.) Then

$$x^{\mathrm{EB}} = E(\lambda|x) = \frac{o+\kappa}{e+1/\theta} = \mu + \frac{\tau^2}{\tau^2+\mu/e}\left(\frac{o}{e}-\mu\right)$$

Proof. First, we work out

$$f(x, \Phi) = \int_0^{+\infty} Po(o, \lambda e)p(\lambda)d\lambda = \int_0^{+\infty} e^{-\lambda E}(\lambda e)^o/o!\,\theta^{-\kappa}\lambda^{\kappa-1}e^{-\lambda/\theta}/\Gamma(\kappa)d\lambda$$

$$= \frac{\theta^{-\kappa}}{\Gamma(o+1)\Gamma(\kappa)} e^{O} \int_{0}^{+\infty} e^{-\lambda(E+1/\theta)} \lambda^{o+\kappa-1} d\lambda$$

$$= \frac{\theta^{-\kappa}}{\Gamma(o+1)\Gamma(\kappa)} e^{O} \int_{0}^{+\infty} e^{-z} z^{o+\kappa-1} (\theta/(1+e\theta))^{o+\kappa-1} d\lambda,$$

(with $z = \lambda(\theta e + 1)/\theta$)

$$= \frac{\theta^{-\kappa}}{\Gamma(o+1)\Gamma(\kappa)} e^{O} \int_{0}^{+\infty} e^{-z} z^{o+\kappa-1} (\theta/(1+e\theta))^{o+\kappa-1} \theta/(1+\theta e) dz$$

(because $\dfrac{dz}{d\lambda} = (\theta e + 1)/\theta$)

$$= \frac{(e\theta+1)^{-\kappa}\Gamma(o+\kappa)}{\Gamma(o+1)\Gamma(\kappa)} (\theta e/(1+\theta e))^{o},$$

because $\int_{0}^{+\infty} e^{-z} z^{x+\kappa-1} dz = \Gamma(x+\kappa)$ by definition of the gamma function. Since $\mu = \kappa\theta$, this latter expression can be written as

$$f(x, \Phi) = f(o, e, \Phi) = \frac{\Gamma(o+\kappa)}{\Gamma(o+1)\Gamma(\kappa)} (\mu e/(\kappa+\mu e))^{o} (1 - \mu e/(\kappa+\mu e))^{\kappa}$$
$$= \frac{\Gamma(o+\kappa)}{\Gamma(o+1)\Gamma(\kappa)} p^{\kappa} (1-p)^{o}$$

which is the *negative binomial* distribution with parameter $p = \mu e/(\kappa+\mu e)$.

Second, we note that

$$x^{EB} = E(\lambda|x) = \int_{0}^{+\infty} \lambda f(\lambda|x) d\lambda = \frac{o+1}{e} f(o+1, e, \Phi)/f(o, e, \Phi)$$

$$= \frac{o+1}{e} \frac{\Gamma(o+1+\kappa)}{\Gamma(o+2)\Gamma(\kappa)} (\mu e/(\kappa+\mu e))^{o+1} (1 + -\mu e/(\kappa+\mu e))^{\kappa}$$

$$\Big/ \left\{ \frac{\Gamma(o+\kappa)}{\Gamma(o+1)\Gamma(\kappa)} (\mu e/(\kappa+\mu e))^{o} (1 - \mu e/(\kappa+\mu e))^{\kappa} \right\}$$

$$= \frac{o+1}{e} \frac{o+\kappa}{o+1} \mu(e/(\kappa+\mu e)) = \frac{o+\kappa}{e+\kappa/\mu} = \frac{o+\kappa}{e+1/\theta},$$

which ends the proof.

In the following we would like to give a demonstration of disease mapping with these empirical Bayes estimates for two data sets. One is on mortality of stomach cancer for nine European countries including 355 regions (Belgium, France (1971–1978); Germany, United Kingdom, Ireland (1976–1980); The Netherlands, Denmark, Luxembourg (1971–1980); and Italy (1975–1979)). The data were provided through the International Agency for Research on Cancer in Lyon. In Figure 7.5 a mortality map is given based on the empirical Bayes estimates (7.5) using the parametric Gamma prior. Evidently, there is large heterogeneity in the map leading to smoothed estimates not very different from the crude SMR-map given in Figure 7.1. Similarly, if a nonparametric prior is used (P estimated nonparametrically), the resulting map (Figure 7.6) is almost identical.

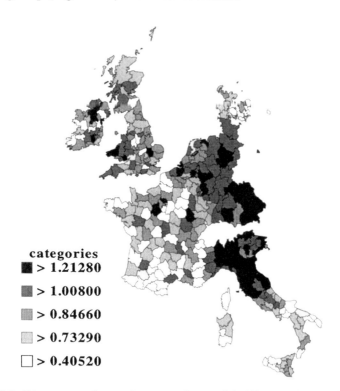

categories
■ > 1.21280
■ > 1.00800
▦ > 0.84660
▨ > 0.73290
□ > 0.40520

Figure 7.5. Disease map of stomach cancer using empirical Bayes estimates of the area SMRs with Gamma prior.

For further case studies using these different mapping procedures, compare also Schlattmann *et al.* (1998) or Kelly (1998).

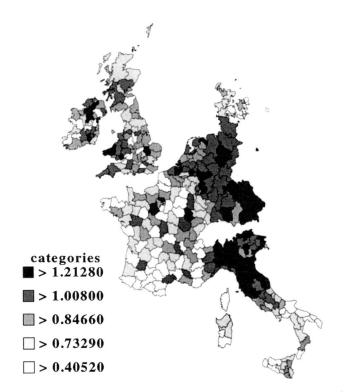

categories
■ > 1.21280
▦ > 1.00800
▨ > 0.84660
□ > 0.73290
□ > 0.40520

Figure 7.6. Disease map of stomach cancer using empirical Bayes estimates of the area SMRs with nonparametric prior.

The second data set is on leukemia mortality for the period 1980–1989 for the counties of the former German Democratic Republic (now the *new states of Germany*). In Figure 7.7 a quintile map of the crude SMRs is given, showing quite a spectrum of variation. The map changes drastically if the empirical Bayes estimate is considered (again using the Poisson-gamma). There is *no* evidence anymore for risk heterogeneity in the map. A similar disease map occurrs if a nonparametric prior is used.

7.5   Estimating map heterogeneity – estimating the prior

We have seen in Section 7.4 that the classical methods (percentile map, significance map) show some *deficiencies* and *disadvantages*. In

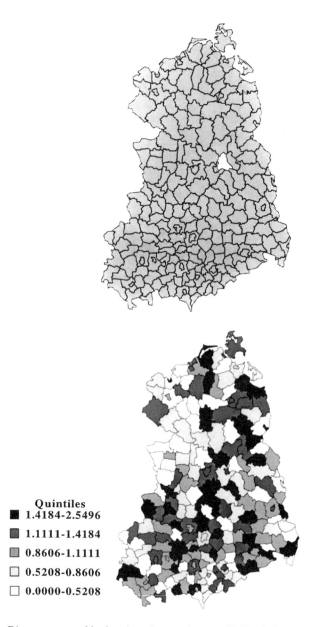

Figure 7.7. Disease maps of leukemia using crude area SMRs (below) and empirical Bayes estimates of the area SMRs with Gamma prior (above); for comparison the map based on empirical Bayes estimates uses the same intervals for their classification into gray colors as the quintile-map of crude SMR-values.

particular, they always force a risk pattern on the map, though this might be an expression of pure random variation. In the previous section it was shown that some of these deficiencies might be improved upon by means of empirical Bayesian methods. However, some problems still remain open.

For the *practical* construction of the disease map it is necessary to use a certain classification system such as: SMR – value → color (gray-pattern). Up to this point we have been still using the percentile method. In addition, the number of colors (gray-patterns) to be used might be in itself of interest. Therefore, we turn now to the *nonparametric prior* itself and use its nonparametric estimate (which is always discrete) as a guide for the choice of color. Assuming again that the mortality cases O_i follow a Poisson distribution with mean $e_i\lambda$, *given the area and* λ, then unconditionally O_i follow a mixture of Poissons

$$O_i \sim \sum_{j=1}^{k} \mathrm{f}(x_i;\lambda_j)p_j = \sum_{j=1}^{k} \mathrm{Po}(o_i;\lambda_j e_i)p_j = \mathrm{f}(o_i, e_i, P),$$

where the mixing distribution P is left nonparametric. As a result we can use the theory of Chapters 2 and 3 to find its estimate $\hat{P}$.

Assuming that the parameter k of the number of components is well estimated, we can use this number to choose the number of colors (gray-patterns) for our map. To demonstrate we look again at the leukemia data of the *New States of Germany* (1980–1989). Here, the NPMLE of P turns out to be the single mass point distribution which puts all its mass at $\lambda = 0.9896$. There is *no* evidence for map heterogeneity and, therefore, we suggest choosing only one color for the map (Figure 7.8). This might be different for other situations. Indeed, if we look at the example of stomach cancer in Europe we find a NPMLE consisting of 7 components.

Likelihood ratio considerations provide evidence that *at least 5 components* appear necessary for an adequate fit of the likelihood (using $2 \times \{\Sigma_i \log \mathrm{f}(o_i, e_i, \hat{P}) - \Sigma_i \log \mathrm{f}(o_i, e_i, \hat{P}_k) \}$ as a criterion; $\hat{P}$ is the NPMLE, whereas $\hat{P}_k$ is the estimate for the fixed components case; see also Chapter 4). Having found an estimate of the number of components it is less clear how to allocate the various areas to the k components. Here, we use the classification rule developed in Chapter 1, Section 1.5. After having computed the posterior density

$$\mathrm{f}(\lambda_j \,|\, o_i, e_i, \hat{P}) = \mathrm{f}(o_i \,|\, \hat{\lambda}_j, e_i)\hat{p}_j/\mathrm{f}(o_i, e_i, \hat{P}), \qquad (7.6)$$

components
☐ p=1.00 l=0.9896

Figure 7.8. Disease map of leukemia in the New States in Germany (1980–1989) using the NPMLE of the heterogeneity distribution.

we can *classify each area i* with SMR-value o_i/e_i into that component (map color) with the *highest posterior probability* (7.6). Note that $f(o_i, e_i, \hat{P}) = \Sigma_j f(o_i \mid \hat{\lambda}_j, e_i)\hat{p}_j$ is just the normalizing constant in Bayes' theorem. Applying this classification rule to the stomach cancer data of Europe we find the disease map as given in Figure 7.9. We can think of this disease map as *adjusted for or cleaned of random variation*. It provides an estimate of the risk structure of cancer for nine European countries. Quite distinguished is the elevated relative risk of about 2 in parts of Bavaria and north-west Italy. It might be of interest to compare this map with the simple disease map of Figure 7.1. It might be argued that much of the evidence in Figure 7.9 could have been seen in Figure 7.1. That might be so. However — leaving aside the issue of the number of gray-patterns — it is one remarkable property of the NPMLE-based map that it provides a much clearer picture of the high risk areas — for example, all the high risk areas (black color) in the northern part of Europe are gone in Figure 7.9. It is one feature

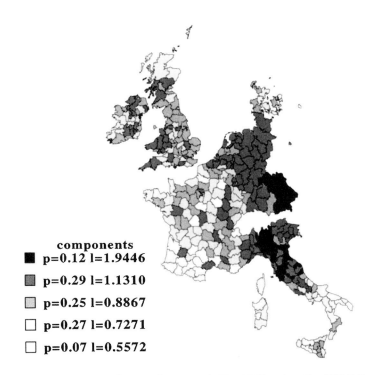

components
- ■ p=0.12 l=1.9446
- ▨ p=0.29 l=1.1310
- ▤ p=0.25 l=0.8867
- □ p=0.27 l=0.7271
- □ p=0.07 l=0.5572

Figure 7.9. Disease map of stomach cancer (1971–1980) using the NPMLE of the heterogeneity distribution.

of the proposed approach that the occurrence of the heterogeneity pattern can be left quite arbitrarily. There is a related approach by Knorr-Held and Raßer (1999) in which it is claimed that the areas of equal risk must be *connected,* which might be a plausible assumption in certain applications whereas in others this might not be appropriate. These ideas can equally be applied for *rate data*: in area i there are o_i cases observed out of N_i at risk. Now, we can either use the binomial kernel in the mixture as it was done in Section 1.6, namely,

$$f(o_i; N_i, P) = \sum_{j=1}^{k} f(o_i; N_i, \lambda_j) p_j = \sum_{j=1}^{k} \binom{N_i}{o_i} \lambda_j^{o_i} (1 - \lambda_j)^{N_i - o_i} p_j$$

or the Poisson approximation leading to

$$f(x_i, \lambda) = Po(o_i, \lambda N_i) = \exp(-\lambda N_i)(\lambda N_i)^{o_i}/o_i! \tag{7.7}$$

where $x_i = o_i/N_i$ is the *observed rate* and λ the *population rate*. We demonstrate this approach with one example discussed in the literature previously.

> **Example 7.4:** Symons *et al.* (1982) discuss the spatial distribution of the SIDS* rate in North Carolina (USA) and suggest a 2-component mixture model (using the Poisson approximation of the binomial mixture kernel). See also Cressie (1993) and Cressie and Chan (1989), the latter using a Markov random field approach. It turns out that there is strong evidence for heterogeneity.** A complete C.A.MAN-analysis with DISMAP shows that the NPMLE is a four-component model. See Figure 7.10. In (a) we see a map using the Poisson kernel (7.7) and for comparison in (b) a map using the binomial kernel. Due to the small number of events (SIDS cases) reltive to the number of live-births, both maps are very similar; just one county is classified differently. Here, the Poisson assumption is less critical. In both cases a four-component mixture model is estimated showing that the number of components is a critical component in the mixture model. The difference in the log-likelihood to the two-component model is about 12.

7.6   Heterogeneity in perinatal mortality for the North West Thames Health Region — data: a case study

In a recent publication, Martuzzi and Hills*** (1995) (see also Martuzzi and Hills 1998) discuss questions of heterogeneity in the true standardized mortality ratios λ of perinatal mortality and suggest modeling this heterogeneity using a gamma distribution, while the distribution of the observed perinatal death is assumed to be a Poisson distribution. The problem of heterogeneity (spatial variation) in geographic epidemiology is of prime interest, since it might indicate that certain areas are at higher risk than others if heterogeneity is present (for an early discussion on heterogeneity aspects in disease maps see Gail 1978). This is the target of many disease atlases (usually mortality) which

---

* SIDS stands for Sudden Infant Death Syndrome.

** It is interesting to note that Cressie and Chan (1989) come to the same result (though the method is different): "The results give an excellent illustration of a phenomenon already well-known in time series, that autocorrelation in data can be due to an undiscovered explanatory variable *(unobserved heterogeneity, D.B.).* Indeed, for 1974–1978 we confirm a dependence of SIDS rate on the proportion of nonwhite babies born" (Cressie and Chan 1989, p. 393).

*** Special thanks go to Dr. Marco Martuzzi who kindly provided the data set for further investigation. Without his cooperation this section would not have been possible.

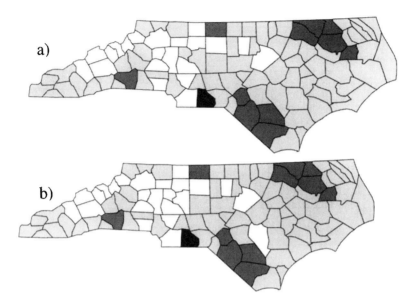

Figure 7.10. Estimated heterogeneity distribution for the SIDS rate in North Carolina (area is county); (a) based upon the approximating Poisson kernel, (b) based upon the Binomial kernel.

are now available for many diseases (cancer, infectious diseases) and for many countries. Martuzzi and Hills (1995) consider a data set concerning the geographic distribution of perinatal mortality in the North West Thames Health Region (NWTHR), England, in the period 1986–1990 on the basis of 515 small area as units for the statistical analysis. The distribution of the SMR is shown in Figure 7.11. This section tries to make the following point. Thus far we have been emphasizing that the heterogeneity analysis is nonparametric in that no assumption for the heterogeneity distribution is assumed. However, the mixing kernel needs to be specified parametrically. This was justified by the argument that the mixing kernel such as the Poisson or binomial is a consequence of the sampling plan if *homogeneity* holds. For SMR-data, however, it is less clear if the conditional distribution of the mortality cases *within one area* follows a Poisson distribution. It is more or less an assumption which could be replaced by other conditional distributions. The consequences in terms of finding heterogeneity might be different. For example, if a Poisson kernel is specified we might find heterogeneity which vanishes when a normal with common variance is assumed. In other words, the mixture kernel can

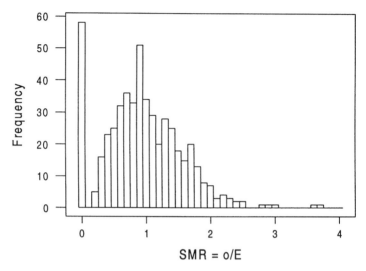

Figure 7.11. Distribution of the Standardized Mortality Ratio for perinatal mortality in the North West Thames Health Region, England, in the period 1986–1990.

be a crucial element in the mixture model, if it is not motivated in a forcing way by the biomedical application. We will demonstrate this by means of the NWTHR-data. For mortality data of SMR-type the Poisson distribution is used as conditional distribution by Martuzzi and Hills (1995) as we have done before as well. Under homogeneity, it is an important property of the Poisson distribution that mean and variance coincide, in other words $E(O_i) = \mathrm{Var}(O_i) = \lambda e_i$. However, in many data applications — as in this one — the empirical variance is larger than the one expected by the Poisson model, a phenomenon called *overdispersion*. In Chapter 4, Section 4.1, we considered over-dispersion tests and have seen that it is possible to consider

$$T = \frac{\frac{1}{n-1}\Sigma_i(O_i + \overline{\mathrm{SMR}}\ e_i)^2/e_i - \overline{\mathrm{SMR}}}{\sqrt{2\overline{\mathrm{SMR}}/(n-1)}},$$

as a test statistic for overdispersion, which is an approximately standard normally distributed test statistic under homogeneity. For the NWTHR data we find: $\overline{\mathrm{SMR}} = 0.95204$ and $(1/(n-1))\Sigma_i (O_i - \overline{\mathrm{SMR}}\ e_i)^2/e_i = 1.0760$, confirming that the SMR data have a larger variance than their mean. Its overdispersion estimate is 0.1240, leading to a value of T = 2.0893 with associated *P*-value of 0.0184.

One possible explanation for the occurrence of overdispersion is a violation of the homogeneity assumption underlying the Poisson distribution. It has been shown already in Example 6.2 (Chapter 6) that population heterogeneity leads to a violation of the mean-variance equality, which is typical for the Poisson model. In particular, if the parameter λ varies with the area, and if this variation can be described by a distribution P, then the situation of (unobserved, latent) heterogeneity has occurred. The marginal distribution of the observed number of death O_i in area i is then given by the density

$$f(o_i) = \int_0^\infty \exp(-\lambda e_i)(\lambda e_i)^{o_i}/o_i! \, p(\lambda) d\lambda \qquad (7.8)$$

as outlined by Martuzzi and Hills (1995). Here p is the probability density function corresponding to the distribution P having mean μ and variance τ^2. Martuzzi and Hills (1995) model a Gamma-distribution for P with mean μ and variance τ^2 (they find estimates for $\mu = 1$ and $\tau^2 = 0.034$). See Figure 7.12. If we leave the distribution P *unspecified* and compute a maximum likelihood estimator of P *nonparametrically* with C.A.MAN, we find for the present data that the nonparametric estimate of heterogeneity consists of just two mass points: $\lambda_1 = 0.7861$ receiving weight $p_1 = 0.4425$ and $\lambda_2 = 1.16$ receiving weight $p_2 = 0.5575$. Again, this kind of approach is often preferred because it is less restrictive in its assumptions concerning the distribution of heterogeneity and it is always as good as the parametric approach can be with respect to goodness of fit. If one computes the log-likelihood difference (times 2) to the distribution putting all its mass at $\lambda = 1$, a value of 6.1 is found which compares favorably to the parametric Gamma with a value of 5.94 (taken from Martuzzi and Hills (1995)). Leaving aside the question of how dramatic the significance found is in both cases, it might be of more importance to note the pattern of both heterogeneity distributions: both are quite *symmetric* (see Figure 7.12). An explanation for this phenomenon can be seen in the fact that the conditional distribution of the SMR given the parameter of the area under consideration has a variance which is too small to model the SMR data adequately. Note that var(O/e) = λ/e, thus fixed by the mean parameter λ. The distribution P of λ in the population takes care of this problem by putting symmetric mass around $\mu = 1$, thus, increasing the unconditional variance of O/e to fit the empirical variance of the data. It will be shown below that an adequate fit of the SMR data can be achieved just by allowing an additional variance parameter in

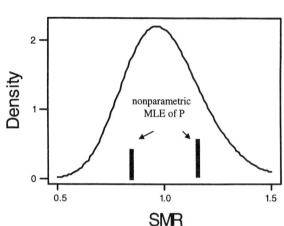

Figure 7.12. Distribution of heterogeneity in the case of parametric gamma distribution with estimated parameters $\mu = 1$ and variance $\tau^2 = 0.034$.

the distributional model for O/e, which can be accomplished by choosing a simple normal density with mean μ and variance τ^2.

The point to be demonstrated is that the SMR distribution is fairly symmetric. Table 7.2 shows selected percentiles of the SMR distribution. Comparisons of the differences of percentiles below the median to the differences of percentiles above the median are fairly symmetric, indicating that a symmetric distributional model would be an alternative way of modeling.

Percentile	Value of percentile
20%	0.4604
30%	0.6302
40%	0.7841
45%	0.8631
50%	0.9144
55%	0.9681
60%	1.0382
70%	1.2203
80%	1.4352

Table 7.2. Selected percentiles in the SMR

Considering the fact that the SMR is a continuous quantity, a simple distributional model for a symmetric distribution is the normal density, which has the advantage of having an extra parameter for the variance, thus relaxing the requirement of binding the variance to the expected value as the Poisson distribution does. This little extra flexibility improves the goodness of fit drastically, as Table 7.3 shows.

Lower	Upper	Observed	Expected	χ^2
Below or at	0.2	61	55.23	0.60
0.2	0.3	13	17.34	1.08
0.3	0.4	19	20.70	0.14
0.4	0.5	20	24.05	0.68
0.5	0.6	26	27.19	0.05
0.6	0.7	37	29.92	1.67
0.7	0.8	38	32.05	1.11
0.8	0.9	37	33.40	0.39
0.9	1.0	43	33.88	2.46
1.0	1.1	31	33.43	0.18
1.1	1.2	31	32.12	0.04
1.2	1.3	25	30.02	0.84
1.3	1.4	25	27.31	0.20
1.4	1.5	16	24.18	2.77
1.5	1.6	17	20.84	0.71
1.6	1.7	20	17.47	0.37
1.7	1.8	13	14.26	0.11
1.8	1.9	15	11.32	1.19
1.9	2.0	9	8.75	0.01
2.1	2.2	1	6.58	4.73
2.2	2.3	8	8.25	0.01
at or above	2.3	10	6.72	1.60
Total				**20.9367**

Chi-square = 20.9367 with 19 d.f. and associated P-value = 0.340307

Table 7.3. Goodness-of-Fit values for the normal distribution with chi-square test for SMR data of perinatal mortality in NWTHR data

This model has one severe disadvantage. Evidently, SMR values below zero cannot occur. One could develop a more formal model for this lower bound of zero by means of a *left-censoring model*, e.g. modifying the density of the SMR as f(smr) = $\exp[-\frac{1}{2}(smr - \lambda)^2/\sigma^2]/\sqrt{2\pi\sigma^2}$ if smr > 0, and F(0) = $\int_{-\infty}^{0} f(s)ds$ if smr $\leq$ 0. This model puts mass

$\int_{-\infty}^{0} f(s)ds$ at 0, which would be conventionally distributed along the line from 0 to $-\infty$ according to the normal density. There is also the possibility of investigating the possibility of heterogeneity with the conditional density being normal and $p(\lambda)$ arbitrary. In other words,

$$f(smr_i) = \int_0^{-\infty} \frac{1}{\sigma}\phi((smr_i - \lambda)/\sigma)p(\lambda)d\lambda.$$

It turns out that the nonparametric maximum likelihood estimator of $p(\lambda)$ does not lead to any significant increase in the likelihood in this case, and the assumption of *homogeneity* can be confirmed. Alternatively, one might stay with the Poisson distribution, but incorporate a modeling element which copes with the *extra zeros*, clearly visible in Figure 7.11. This can be accomplished by means of

$$(1 - p) \text{ Po}(o_i, 0, e_i) + p\text{Po}(o_i, \lambda, e_i) \tag{7.9}$$

where $\text{Po}(o_i, 0, e_i) = 1$, if $o_i = 0$, and $= 0$ otherwise. A C.A.MAN fit shows that weight $\hat{p} = 0.99$ is given to $\hat{\lambda} = 1.0017$, and almost no elevated risks are left. Model (7.9) is also called the *zero-inflated Poisson* model and has recently experienced more attention (Lambert 1992; Dietz and Böhning 1997; Böhning *et al.* 1999).

This analysis shows that sometimes an equally good fitting model can be achieved. The question arises: how much does it matter if two models are found that provide almost similar fits? From a purely statistical point it might be argued that there are two equally good fitting models, and only a good (or reasonably well) fitting model is replaced by another one with similar fitting properties. However, from an epidemiological point of view the causes and consequences in both models are rather different. In the heterogeneity model (and here it *does not matter* if it is the parametric gamma or nonparametric heterogeneity distribution) spatial variation in risk is allowed and modeled, although only indirectly is support via the likelihood ratio test (hypothesis of homogeneity against hypothesis of heterogeneity) provided. Having found significant heterogeneity the conclusion is drawn that there are areas with increased and areas with decreased relative risk, and naturally we attempt to identify those. In the second model (and equally in the third, the *zero-inflated Poisson* model), which simply fits a distribution with mean value* near 1, the conclusion is drawn that there is *no* heterogeneity in relative risk, and there is no need

---

* For the ZIP-model this refers to the second component mean.

for further investigation and search for areas with increased health risk and their potential causes. The reason for this is as follows.

Heterogeneity found in epidemiological data of the SMR type can have various causes. One cause is that there is a population variation of the theoretical parameter of the SMR. Or, it might be that the conditional distribution (here the observed deaths) given the area's specific parameter is *unable* to capture the amount of variation *within the area* under consideration, in other words it is *erroneously specified*. It is believed that this is the case in the NWTHR data if the Poisson distribution as conditional distribution is used. Therefore, the found heterogeneity is *artificial* and vanishes completely if a conditional distribution is used with an additional variance parameter. This conditional distribution need not necessarily be the normal, though it was found sufficient in the present situation. Alternatively, one could think of the negative binomial distribution* with mean parameter μ and variance τ^2 as a candidate distribution (among others) for the conditional distribution of perinatal death cases given the area, and then look for further heterogeneity in the mean parameter μ. It appears to be a central issue that the conditional distribution of the SMR_i in area i given the true parameter λ_i in area i is chosen correctly. If not, there is the risk of forcing a heterogeneity distribution on the parameters λ_i to fit the observed data variability.

The contribution of this section highlights the old wisdom that a phenomenon or fact can have a variety or diversity of explanations. Overdispersion can occur because of heterogeneity of the underlying true parameter which can be captured by parametric (such as the gamma-poisson) or nonparametric (such as the nonparametric Poisson) mixture models. But it can also occur because the Poisson assumption for the *conditional* distribution is violated.

7.7   Space-time mixture modeling

Recently, increasing interest has developed in disease mapping when there is an additional time component (see Bernardinelli *et al.* 1995). In the framework of observed and expected deaths this might be formulated as follows. Let O_{it} (e_{it}) be the observed (expected) number of death cases at time period t and area i, where $t = 1, ..., T$ and $i = 1, ..., n$. It is straightforward to apply mixture modeling to each *time*

---

* The negative binomial distribution might be more appropriate for a count distribution than the normal.

period separately, leading to T different mixture models, for each time
period one. The associated parameters might be simply indexed as $\lambda_j^{(t)}$
for the jth component mean in period t, and $p_j^{(t)}$ for the jth component
weight in period t. The problem with this approach lies in the fact that
maps (mixture models) over time might be difficult to compare, since
the same area might not only be in different components at different
times, but also if they were in the same component, their means might
be different.

Example 7.5: To give a demonstration we come back to the SIDS data
of North Carolina which have been discussed in Section 7.5, Example
7.4, previously. SIDS data are not only available for the period from
1974–1978, but also for the next 5-year period 1979–1984. See also
Cressie (1993). The results of fitting to separate mixture models are
displayed in Table 7.4.

$\lambda_j^{(1)}$	$p_j^{(1)}$	$\lambda_j^{(2)}$	$p_j^{(2)}$
0.0090	0.01	0.0054	0.01
0.0037	0.16	0.0029	0.23
0.0021	0.52	0.0020	0.52
0.0012	0.31	0.0013	0.24

Table 7.4. Results of fitting two separate mixture model to the 2 periods
SIDS data of North Carolina; $t = 1$: 1974–1978; $t = 2$: 1979–1984

Note that in Table 7.4 both mixture models have the same number of
components ($k = 4$), which is not necessarily the case. This makes the
comparison easier. Note also that the third and fourth component means
are rather close, whereas the first and second component means do not
agree. Obviously, they are lower for the second period. Therefore, a
different gray shading has been used for these two components (see
Figure 7.13).

An alternative approach tries to look at *heterogeneity in space and
time simultaneously.* Suppose there exist k *time-space* components
behind the data, e.g., there exist k *clusters* or *components* $j = 1, \ldots k$
such that all areas i_j at times t_j belong to the component j. Let again
Z_j denote the *unobserved* indicator variable, which takes on the value
1, if x is from sub-population j, and is 0 otherwise. Let $z_{ij}^{(t)}$ denote the
value of Z_j for observation $x_i^{(t)} = o_i^{(t)}/e_i^{(t)}$ at time period t. The likelihood
for the pair $(x_i^{(t)}, z_{ij}^{(t)})^T$ is

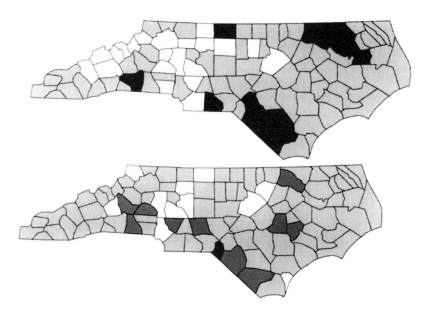

Figure 7.13. Separately estimated heterogeneity distribution for the SIDS-rate in North Carolina (area is county) for the period 1974–1978 (above) and the period 1979–1984 (below); maps use different gray scale

$$\text{Pr}(X_i^{(t)} = x_i^{(t)}, \; Z_{i1}^{(t)} = z_{i1}^{(t)}, ..., Z_{ik}^{(t)} = z_{ik}^{(t)})$$

$$= \text{Pr}(X_i^{(t)} = x_i^{(t)} \mid Z_{i1}^{(t)} = z_{i1}^{(t)}, ..., Z_{ik}^{(t)} = z_{ik}^{(t)}) \; \text{Pr}(Z_{i1}^{(t)} = z_{i1}^{(t)}, ..., Z_{ik}^{(t)} = z_{ik}^{(t)})$$

$$= \prod_{j=1}^{k} \text{f}(x_i^{(t)}, \lambda_j)^{z_{ij}^{(t)}} p_j^{z_{ij}^{(t)}},$$

and correspondingly for the *full* likelihood

$$\prod_{t=1}^{T} \prod_{i=1}^{n} \prod_{j=1}^{k} \text{f}(x_i^{(t)}, \lambda_j)^{z_{ij}^{(t)}} p_j^{z_{ij}^{(t)}},$$

assuming *independence conditional on the knowledge of the assignment indicator* $z_{ij}^{(t)}$. Using the connection between complete and incomplete likelihood (see Chapter 3, Section 3.5) we achieve the familiar marginal or *mixture* log-likelihood

$$\sum_{t=1}^{T} \sum_{i=1}^{n} \log \left\{ \sum_{j=1}^{k} p_j \text{f}(x_i^{(t)}, \lambda_j) \right\}. \tag{7.10}$$

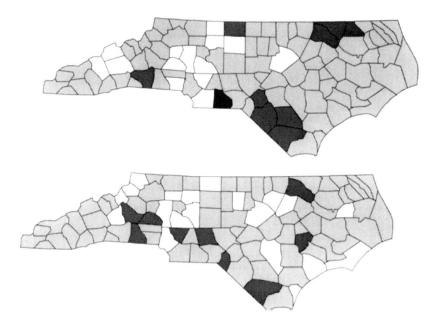

Figure 7.14. Jointly estimated heterogeneity distribution for the SIDS rate in North Carolina (area is county) for the period 1974–1978 (above) and the period 1979–1984 (below); both maps use identical gray scale.

We work with this mixture likelihood as we have done before. Note, however, that each area is classified T times when using the maximum posterior probability as the allocation rule. For example, it might occur that an area is classified at different times into different space-time components.

Example 7.6: We continue the discussion of the SIDS data of North Carolina from Example 7.5. We will try to find one mixture model which explains both variations, in time and space. The results of the C.A.MAN analysis are provided in Table 7.5. In this case, we find again a four-component model. Note that the number of parameters in the model is quite reduced in comparison to the model used previously in Example 7.5. Instead of $T \times (2k - 1)$ we have only $2k - 1$ parameters in this model. We note that 62 counties do not change their allocation to the space-time components, whereas 38 counties do so, most prominently the single high risk county *Anson*, which was in the high risk group (8.5 deaths per 1000 live-births). It changes its allocation to the next lower component (3.4 deaths per 1000 live-births). In addition, there are more movements from higher risk components in the '74–78 period to lower

λ_j	p_j
0.0085	0.01
0.0034	0.15
0.0021	0.54
0.0013	0.30

Table 7.5. Results of fitting one common mixture model to the 2 periods SIDS data of North Carolina; $t = 1$: 1974–1978; $t = 2$: 1979–1984

risk components in the '79–84 period, namely 23, then vice versa, namely 15, indicating a general, decreasing time trend in SIDS rate. The overall pooled SIDS death-rate changes from 2.021 deaths per 1000 live-births in the '74–78 period to the 1.979 deaths per 1000 live-births. Just this last consideration might point in the direction of modeling changes in time by *hidden Markov chain* models.

CHAPTER 8

Various C.A.MAN applications

In this chapter we have collected a variety of different applications of mixture models. We begin with the modeling of heterogeneity in fertility studies.

8.1   Fertility studies

In fecundability studies (see Ridout and Morgan 1991) the situation is as follows. If X represents the cycle number in which pregnancy is reached, then X follows the *geometric distribution* with density $f(x, \lambda)$ $= (1 - \lambda)^{x-1}\lambda$, for $x = 1, 2, 3, \ldots$. The parameter λ is called the *fertility parameter*. It indicates the risk for pregnancy. In these studies, we have to cope with the problem of *censoring*: in the study period no pregnancy might occur. If x denotes the last observed cycle, we ask for the probability that pregnancy occurs in some later cycle: $\Pr\{X > x\}$ $= \sum_{y=x+1}^{\infty} \lambda(1 - \lambda)^{y-1} = \lambda(1 - \lambda)^x \sum_{y=0}^{\infty} (1 - \lambda)^y = \lambda(1 - \lambda)^x/(1 - (1 - \lambda))$ $= (1 - \lambda)^x$. Therefore, $f(x, z, \lambda) = (1 - \lambda)^x$ if x is censored ($z = 1$), or $f(x, z, \lambda) = (1 - \lambda)^{x-1}\lambda$ if x is *not* censored ($z = 0$). In other words,

$$f(x, z, \lambda) = (1 - \lambda)^{xz} (1 - \lambda)^{(x-1)(1-z)} \lambda^{(1-z)} \tag{8.1}$$

To demonstrate the modeling we look at a data set originally discussed by Weinberg and Gladen (1986) and later by Ridout and Morgan (1991). See Table 8.1. Of the total of 486 couples 12 remained unpregnant at the end of the study period. The MLE is in this case $1/\hat{\lambda} = 3.455$ with a log-likelihood of -1336.26. If we allow for heterogeneity, corresponding to (1.1) we achieve the nonparametric geometric mixture

$$f(x, z, P) = \sum_{j=1}^{k} f(x;z;\lambda_j)p_j \tag{8.2}$$

Cycle	1	2	3	4	5	6	7	8	9	10	11	12	>12
# pregnancies	198	107	55	38	18	22	7	9	5	3	6	6	12

Table 8.1. Observed cycles to pregnancy (nonsmokers), data according to Weinberg and Gladen (1986)

Fitting the model (8.2) with C.A.MAN we find a two-component structure for heterogeneity (k = 2): the NPMLE consists of two support points, namely, $\hat{\lambda}_1$ = 1 and $1/\hat{\lambda}_2$ = 3.632419 with associated weights $\hat{p}_1$ = 0.0418 and $\hat{p}_2$ = 0.9582, respectively. The log-likelihood is −1335.08, which compares to the homogeneous log-likelihood nonsignificantly. Basically, the result is that we have a *homogeneous* population of couples (the likelihood ratio test is borderline if the NPMLE is compared to the homogenous MLE). However, there is some inflation of couples with success in the first cycle (4%). For the homogeneous population the fertility parameter is 0.2894.

A second group of 1274 couples is presented in Table 8.2. They have in common that the women were using the contraceptive pill before trying to become pregnant.

Cycle	1	2	3	4	5	6	7	8	9	10	11	12	>12
# pregnancies	383	267	209	86	49	122	23	30	14	11	2	43	35

Table 8.2. Observed cycles to pregnancy (contraceptive pill users), data according to Harlap and Baras (1984)

It turns out that this population is more heterogeneous than the one considered before. The homogeneity MLE is $1/\hat{\lambda}$ = 3.5 with a log-likelihood of −2635.939. Analyzing potential heterogeneity with C.A.MAN suggests that the population can be partitioned into two equally sized subpopulations with parameters $1/\hat{\lambda}_1$ = 2.4 and $1/\hat{\lambda}_2$ = 4.8. The C.A.MAN-results are in detail: the NPMLE consists of two support points, namely, $1/\hat{\lambda}_1$ = 2.4044 and $1/\hat{\lambda}_2$ = 4.8462 with associated weights $\hat{p}_1$ = 0.4971 and $\hat{p}_2$ = 0.5029, respectively. The log-likelihood is −2625.218, which means a difference to the homogeneity log-likelihood of about 10. This difference supports the theory that we are dealing with a heterogeneous population (for details on the likelihood ratio test in mixture models, see Chapter 4) consisting of two subpopulations of about almost equal size with fertility parameter $\hat{\lambda}_1$ = 0.42 and $\hat{\lambda}_2$ = 0.21, respectively.

It should be pointed out that in the approach given here no *parametric* form of the mixing distribution is used. In this respect it differs from other models in which certain parametric distributional forms on λ are assumed, such as the beta-distribution (see Ridout and Morgan (1991) for comparison).

8.2 Modeling a diagnostic situation

We consider a classical mixture problem: the situation of medical diagnosis. Some observable measurement X (test result) is available, though of interest is the disease status D, a binary variable (1/0) which is not observable in a direct way. The distribution of the test result X might be known for the diseased and non-diseased population, though the disease occurrence distribution is usually unknown. In statistical terms, $f(x \mid d)$ is available for $d = 0$, 1, but the joint density $f(x, d)$ is unknown. However, the marginal density can be computed as

$$f(x) = f(x \mid D = 1)p + f(x \mid D = 0)(1 - p) \qquad (8.3)$$

with $p = \Pr(D = 1)$, where the margin is taken over the disease status values. Evidently, $f(x)$ is a mixture distribution in which the component distributions are case and non-case distributions and the mixing weights are provided by the prevalence rate. The prevalence rate p, as well as the parameters of the component densities, has to be estimated from data. For an illustration we look at the data given in Figure 8.1. It consists of systolic blood pressure measurements of 495 mine workers in Ghana. Obviously, the distribution is not symmetric and one might be interested to find out if there is a second, high risk group in the population. A suitable component density model is the normal, e.g., $f(x \mid D = 1) = \varphi((x - \lambda_2)/\sigma)/\sigma$ and $f(x/D = 0) = \varphi((x - \lambda_1)/\sigma)/\sigma$. After estimating the three parameters λ_1, λ_2, and p, the patients can be classified *into the two* disease groups using Bayes' rule:

$$\Pr(D = 1 \mid X = x) = \frac{f(x \mid D = 1)p}{\{f(x \mid D = 1)p + f(x \mid D = 0)(1 - p)\}} \qquad (8.4)$$
$$= f(x \mid D = 1) \times p / f(x).$$

Similarly, we find that $\Pr(D = 0 \mid X = x) = f(x \mid D = 0) \times (1 - p) / f(x)$. A person is classified into the group with the highest posterior probability, or in this case, into the *diseased* group if $\Pr(D = 1 \mid X = x) > 1/2$. For the blood pressure data set Table 8.3 provides the results of the

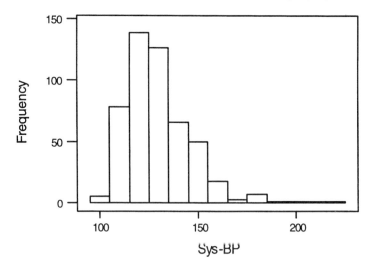

Figure 8.1. Histogram of systolic blood pressure for a sample from a population of mine workers in Ghana (data from Gunga *et al.* 1991).

C.A.MAN output. With these parameter values the posterior distribution can be computed and the patients can be classified into the two groups accordingly.

λ_1	λ_2	p	σ^2
127.478	182.342	0.0281	185.373

Table 8.3. Estimates for the 4 parameters of the mixture model (8.3) with normal component density

There is an alternative to the general classification rule given above. It is based on the idea of *unbiased* estimation of prevalence. The cut-off value should be chosen according to

$$p \int_{-\infty}^{c} \varphi((x-\lambda_2)/\sigma)\,\mathrm{d}x = (1-p)\int_{c}^{\infty} \varphi((x-\lambda_1)/\sigma)\,\mathrm{d}x \qquad (8.5)$$

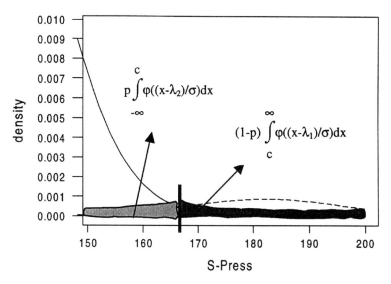

Figure 8.2. Illustration of the construction of the cut-off value according to (8.5).

This cut-off value is recommended in Titterington *et al.* (1985) (for an illustration see Figure 8.2) and is in contrast to a cut-off value according to the intersection of the two curves: $p\,\varphi((c - \lambda_2)/\sigma) = (1 - p)\,\varphi((c - \lambda_1)/\sigma)$. The latter can be given in closed form as $c = \{\sigma^2 \log(p/(1 - p)) + \tfrac{1}{2}\lambda_1^2 - \tfrac{1}{2}\lambda_2^2\}\,/(\lambda_1 - \lambda_2)$, which is not possible for the cut-off value determined by the equation (8.5). A justification of the cut-off value (8.5) is given in the following theorem.

Theorem 8.1: Let D_c be the classification rule based on the cut-off value c given (8.5): $D_c = 1$ if $X > c$, and $D_c = 0$ if $X \le c$. Then:

$$\Pr\,(D_c = 1) = p$$

Proof.

$$\Pr(D_c = 1) = \Pr(X > c) = p\int_c^\infty \varphi((x - \lambda_2)/\sigma)\,dx + (1 - p)\int_c^\infty \varphi((x - \lambda_1)/\sigma)\,dx$$

$$= p\left(1 - \int_{-\infty}^c \varphi((x - \lambda_2)/\sigma)\,dx\right) + (1 - p)\int_c^\infty \varphi((x - \lambda_1)/\sigma)\,dx$$

$$= p - (1 - p)\int_c^\infty \varphi((x - \lambda_1)/\sigma)\,dx + (1 - p)\int_c^\infty \varphi((x - \lambda_1)/\sigma)\,dx$$

where we have used in the last equation that $p \int_{-\infty}^{c} \varphi((x-\lambda_2)/\sigma)dx =$ $(1-p)\int_{c}^{\infty}\varphi((x-\lambda_1)/\sigma)dx$ according to (8.5). This ends the proof.

C.A.MAN is capable of computing this optimal threshold-value (8.5). For the blood pressure data this value is found to be $c = 164.9962$. The marginal density with this optimal cut-off value is provided in Figure 8.3.

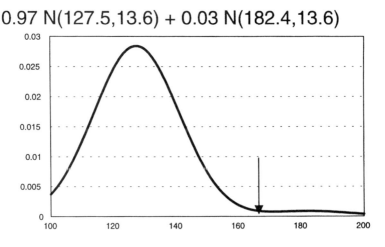

0.97 N(127.5,13.6) + 0.03 N(182.4,13.6)

Figure 8.3. Marginal density f(x) with cut-off value according to (8.5).

8.3   Interval-censored survival data

We now look at a situation which is rather different in the way the mixture setting occurs. The situation (and notation) is as in Gentleman and Geyer (1994). We consider the time X until a certain life event (death, disease occurrence) appears. However, for certain reasons the *survival time* X is not observed completely. Instead, the unobserved survival time X (with distribution F) is supposed to lie in an observed open interval (L, R), corresponding to the last inspection time prior to the life event and the first inspection time after the life event. Thus, the observable data are open intervals $I_1, \ldots, I_n$, with $I_i = (L_i, R_i)$ for $i = 1, \ldots, n$. According to Gentleman and Geyer (1994), the likelihood is defined as $\prod_{i=1}^{n}\{F(R_i-)-F(L_i)\}$. A general introduction to nonparametric maximum likelihood estimation for censored data can be found in Groeneboom and Wellner (1992). If $\{s_j\}_{j=1}^{m}$ denotes the *unique ordered* elements of $\{0, \{L_i\}_{i=1}^{n}, \{R_i\}_{i=1}^{n}\}$ and α_{ij} the indicator of the event $(s_{j-1}, s_j) \subseteq I_i$, $p_j = F(s_j) - F(s_{j-1})$, then the likelihood can be written as

$$L = \prod_{i=1}^{n}\left(\sum_{j=1}^{m}\alpha_{ij}p_j\right). \qquad (8.6)$$

Let $\eta_i = (\alpha_{i1}p_1 + \ldots + \alpha_{im}p_m)$ be the ith contribution to the likelihood. Finding the maximum likelihood of L becomes the problem of maximizing $l = \log L$ in $p_1, \ldots, p_m$ subject to $p_j \geq 0$ for $j = 1, \ldots, m$ and $p_1 + \ldots + p_m = 1$. The problem is to maximize $l(\boldsymbol{p})$ in the finite dimensional probability simplex $\Delta = \{\boldsymbol{p} = p_1\boldsymbol{e}_1 + \ldots + p_m\boldsymbol{e}_m \mid p_j \geq 0$ for $j = 1, \ldots, m$ and $p_1 + \ldots + p_m = 1\}$ with $\boldsymbol{e}_j$ being the vector having only 0s and exactly one 1 at the jth position (*vertex*).

The purpose of this section is to point out some analogies of the problem at hand and the problem of finding the nonparametric maximum likelihood estimator of a mixing distribution, and what can be gained from exploiting this analogy. A theory for the nonparametric maximum likelihood estimator of the mixing distribution P giving weights $p_1, \ldots, p_m$ to parameter values $\lambda_1, \ldots, \lambda_m$ for some parametric density $f(x, \lambda)$ has been developed in the previous chapters. The problem of finding the maximum likelihood estimator of the mixing distribution is to maximize

$$L^* = \prod_{i=1}^{n}\left\{\sum_{j=1}^{m}f(x_i;\lambda_j)p_j\right\}$$

in $p_1, \ldots, p_m$ if $\lambda_1, \ldots, \lambda_m$ are considered to be known. This clearly points the analogy to (8.6). We restate the relevant theorems.

Characterizing the maximum likelihood estimate. As pointed out by Gentleman and Geyer (1994) $l(\boldsymbol{p})$ is a concave function on Δ. Define the *directional derivative* $\Phi(\boldsymbol{p}, \boldsymbol{q}) = \lim_{\alpha\to0}(l((1-\alpha)\boldsymbol{p} + \alpha\boldsymbol{q}) - l(\boldsymbol{p}))/\alpha$ as before for any $\boldsymbol{p}$ in Δ and any direction $\boldsymbol{q}$. Note that $\Phi(\boldsymbol{p}, \boldsymbol{q}) = \nabla l(\boldsymbol{p})^{\mathrm{T}}\boldsymbol{q} - n$ is a linear function of the direction $\boldsymbol{q}$. Here $\nabla l(\boldsymbol{p})^{\mathrm{T}} = (d_1, \ldots, d_m)^{\mathrm{T}}$ with $d_k = \partial l/\partial p_k = \sum_{i=1}^{n}(\alpha_{ik}/\eta_i)$. As Gentleman and Geyer (1994) point out, d_k is the sum of $1/\eta_i$ for all individuals whose intervals, I_i, intersect the interval (s_{k-1}, s_k). Because l is concave $\Phi(\boldsymbol{p}, \boldsymbol{q}) \geq l(\boldsymbol{q}) - l(\boldsymbol{p})$ and hence,

$$\sup_{q \in \Delta} \Phi(p, q) \geq l(\hat{p}) - l(p). \tag{8.7}$$

Since $\Phi(p, q) = \sum_{j=1}^{m} q_j \Phi(p, e_j)$, where e_j is the jth vertex of Δ, the inequality (8.7) becomes

$$D(p) = \sup_{1 \leq k \leq m} \sum_{i=1}^{n} \frac{\alpha_{ik}}{\eta_i} - n \geq l(\hat{p}) - l(p) \tag{8.8}$$

Clearly, if $D(\hat{p}) = 0$, $\hat{p}$ maximizes l globally. This is a special form of the *general mixture maximum likelihood theorem* (Theorem 2.1), stating that $D(\hat{p}) = 0$, if and only if $\hat{p}$ maximizes l globally in Δ. Moreover, if $\hat{p}_k > 0$, then $\Phi(\hat{p}, e_k) = 0$, in other words $\sum_{i=1}^{n} (\alpha_{ik}/\eta_i) = n$. Frequently, it is important to consider the likelihood in its geometric mean version where (8.6) becomes $(\eta_1 \times \ldots \times \eta_n)^{1/n}$. Then $\bar{l} = (1/n)l$, $\bar{d}_k = \partial \bar{l}/\partial p_k = \sum_{i=1}^{n} (\alpha_{ik}/\eta_i)/n$ and, the *general mixture maximum likelihood theorem* takes the form $\bar{d}_k \leq 1$ for all $k = 1, \ldots, m$ and $\bar{d}_k = 1$ for those k with p_k strictly positive. We demonstrate this theorem at the example given in Gentleman and Geyer (1994, Section 4)

Example 8.1: Let the data consist of the six intervals (0, 1], (1, 3], (1, 3], (0, 2], (0, 2], (2, 3]. This leads to the following 6×3 matrix of indicators

$$A = (\alpha_{ij}) = \begin{pmatrix} 1 & 0 & 0 \\ 0 & 1 & 1 \\ 0 & 1 & 1 \\ 1 & 1 & 0 \\ 1 & 1 & 0 \\ 0 & 0 & 1 \end{pmatrix}$$

and the likelihood is $L = \prod_{i=1}^{n} (\sum_{j=1}^{m} \alpha_{ij} p_j) = \prod_{i=1}^{6} (\alpha_{i1} p_1 + \alpha_{i2} p_2 + \alpha_{i3} p_3)$. We have that $d_1 = 1/p_1 + 2/(p_1 + p_2)$, $d_2 = 2/(p_2 + p_3) + 2/(p_1 + p_2)$, $d_3 = 2/(p_2 + p_3) + 1/p_3$. From this it is clear that $p = (1/3, 1/3, 1/3)^T$ is the maximum likelihood estimate, since $d = (6, 6, 6)^T$ is meeting the condition of the mixture maximum likelihood theorem. The point $(1/2, 0, 1/2)^T$ is clearly *not* the maximum likelihood estimate, since the partial derivative $d_2 = 8 > 6$ at this point. Thus, the likelihood can be increased by moving into the direction of the second vertex e_2. This can be seen also by looking at various plots of the likelihood when it is written as a function of p_1 and p_2 alone:

$$\prod_{i=1}^{6} \{\alpha_{i1}p_1 + \alpha_{i2}p_2 + \alpha_{i3}(1 - p_1 - p_2)\} = p_1(1 - p_1)^2(p_1 + p_2)^2(1 - p_1 - p_2).$$

The log-likelihood is in this case

$$l(p) = \log(p_1) + 2 \log(1 - p_1) + 2 \log(p_1 + p_2) + \log(1 - p_1 - p_2).$$

Figure 8.4 shows a three-dimensional graph of its surface. It is a very regular, concave surface, though rather flat. Since it is not completely clear from Figure 8.4 where the maximum value is attained on this surface, we provide a contour plot in Figure 8.5, which clearly encircles the maximum likelihood estimate $(1/3, 1/3, 1/3)^{\mathrm{T}}$.

Example 8.2: The second example is more challenging. The data are from Finkelstein and Wolfe (1985) and consist of time intervals in which cosmetic deterioration for early breast cancer patients treated with radiotherapy occurred in 46 individuals. The intervals are listed in Table 1 in Gentleman and Geyer (1994). To avoid repetition of the data, only the corresponding indicator 48×14-matrix $A = (\alpha_{ij})$, which is not provided in Gentleman and Geyer (1994), is given instead:

```
0 0 0 0 0 0 0 0 0 0 0 0 0 1
0 1 1 0 0 0 0 0 0 0 0 0 0 0
1 1 0 0 0 0 0 0 0 0 0 0 0 0
0 0 0 0 0 0 0 0 0 0 0 0 0 1
0 0 0 0 0 0 0 0 0 0 0 0 0 1
0 0 1 1 1 0 0 0 0 0 0 0 0 0
0 0 0 0 0 1 1 1 1 1 1 1 1 1
0 0 1 1 0 0 0 0 0 0 0 0 0 0
0 0 0 0 0 0 0 0 0 0 0 1 1 0
1 1 1 0 0 0 0 0 0 0 0 0 0 0
1 1 1 0 0 0 0 0 0 0 0 0 0 0
0 0 0 0 1 1 1 1 1 1 1 1 1 1
0 0 0 1 0 0 0 0 0 0 0 0 0 0
0 0 0 0 0 0 1 1 1 1 1 1 1 1
0 0 0 0 0 0 0 0 0 0 0 0 0 1
0 0 0 0 0 0 0 0 0 0 0 0 0 1
0 0 0 0 0 0 0 1 1 1 1 0 0 0
0 0 0 0 0 0 0 0 0 0 0 0 0 1
0 0 0 0 0 0 0 0 1 1 1 1 0 0
0 0 0 0 0 0 0 0 0 0 0 0 0 1
0 0 0 0 0 0 0 0 1 0 0 0 0 0
0 0 0 0 0 0 0 0 0 0 1 1 1 0
```

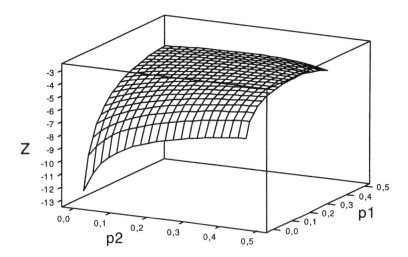

a)

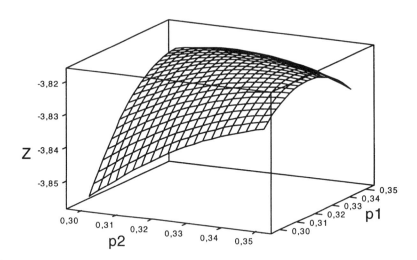

b)

Figure 8.4. Graph of the log-likelihood (Z), $\log(p_1) + 2\log(1 - p_1) + 2\log(p_1 + p_2) + \log(1 - p_1 - p_2)$, for Example 8.1; (a) p_1 and p_2 ranging from 0 to 0.5, (b) p_1 and p_2 ranging from 0.30 to 0.35.

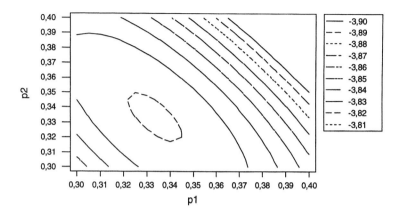

Figure 8.5. Contour plot of the log-likelihood (Z), $\log(p_1) + 2\log(1 - p_1) + 2\log(p_1 + p_2)$ + $\log(1 - p_1 - p_2)$, for Example 8.1.

```
0 0 0 0 0 0 0 0 0 0 0 0 0 1
0 0 0 0 0 0 0 0 0 0 1 1 1 1
0 0 0 0 0 0 0 0 0 0 0 1 1 1
0 0 0 0 0 0 0 0 0 0 0 0 1 1
0 0 0 0 0 1 1 0 0 0 0 0 0 0
0 0 0 0 0 0 0 0 0 0 0 0 0 1
0 0 0 1 1 1 0 0 0 0 0 0 0 0
0 0 0 0 0 0 0 0 0 0 0 1 1 1
0 1 1 1 0 0 0 0 0 0 0 0 0 0
0 0 0 0 0 0 0 0 0 0 0 1 1 1
1 0 0 0 0 0 0 0 0 0 0 0 0 0
0 0 0 0 0 0 1 1 1 1 1 1 1 1
0 0 0 0 0 0 1 1 1 1 1 1 1 1
0 0 0 0 0 0 0 0 0 0 1 1 1 1
0 1 1 0 0 0 0 0 0 0 0 0 0 0
0 0 0 0 0 0 1 1 1 1 0 0 0 0
0 0 0 0 0 1 1 0 0 0 0 0 0 0
0 0 0 0 0 0 1 1 1 1 1 1 1 1
0 0 0 0 0 0 0 0 1 1 1 1 1 1
0 0 0 0 0 0 0 0 1 1 1 1 1 1
0 0 0 0 0 0 1 1 0 0 0 0 0 0
0 0 0 0 0 0 0 0 0 0 0 1 1 1
0 0 0 0 0 0 0 0 0 1 1 1 1 1
0 0 0 0 0 0 0 0 0 0 1 1 1 1
```

For these data, it is not possible to find the maximum likelihood estimate in a closed form expression. Algorithmic approaches have to be applied.

Algorithms. A simple and reliable algorithm is a version of the EM algorithm (Dempster, Laird and Rubin 1977) which in this case reduces to the iteration p^{EM} with jth component $p_j^{EM} = \overline{d}_j p_j$. The EM algorithm has the celebrated monotonicity property $l(p^{EM}) \geq l(p)$, if p is the current iterate, which ensures reliable convergence to the global maximum for arbitrary starting values with components all strictly positive. However, for flat likelihood surfaces — as in this case — its convergence can be rather slow. In addition, many of the p_j might be exactly zero, and — as pointed out by Gentleman and Geyer (1994) — the EM algorithm takes extra time to identify these (in Example 2 there are 6 of 14 weights zero). Gentleman and Geyer (1994) suggest the strategy to restart the EM algorithm with the candidate weights set to zero. This will speed up the convergence, but it will also run the risk of *falsely* setting a weight to zero (such as p_2 in Table 2 of Gentleman and Geyer 1994). Since the EM algorithm never changes zero weights, it will only maximize the log-likelihood in the corresponding sub-simplex. Indeed, if we consider p given in column 7 of Table 2 in Gentleman and Geyer (1994), that is $p_2 = 0$ in particular, we find that $\overline{d}_2 = 1.1327$, indicating that the likelihood can be increased by putting more weight to the second vertex. For mixture models a variety of alternative algorithms have been presented in Chapter 3, the most useful one being the *vertex-exchange algorithm* (VEM). For this application it takes the form of the iteration $p + \alpha \, p_{\min} \, (e_{\max} - e_{\min})$, max and min are integer indices between 1 and m, $p_{\min}$ corresponds to the weight of the vertex $e_{\min}$, and α is a monotonic or optimal step-length in the closed interval [0, 1] (see Chapter 3 for details). Since in this context the possibility of many weights being zero arises, the VEM should be quite useful here as well. The index "max" can be determined as $d_{\max} = \max \{d_j \mid j = 1, ..., m\}$, whereas "min" can be found as that index for which the partial derivative is minimized under those with positive weight p_j: $d_{\min} = \min \{d_j \mid j = 1, ..., m$ and $p_j > 0\}$. The EM algorithm, the VE-Method, among other algorithms has been included in the statistical package for mixture analysis C.A.MAN (Böhning, Schlattmann, and Lindsay 1992; Böhning, Dietz, and Schlattmann 1998). This package computes the mixing distribution for various densities, including the option of the *known density* case (Titterington, Smith, and Makov 1985, p. 152). If one uses this option, e.g., if one thinks of the matrix

$A = (\alpha_{ij})$ as $F = (f(x_i, \lambda_j))$, with $\lambda_1, \ldots, \lambda_m$ known, then C.A.MAN can be used directly without further modification, and the whole variety of powerful algorithms is available for this setting of interval censored data.

Example 8.2 (continued): The maximum likelihood estimate given in Table 8.4 could be identified using any of the algorithms available in C.A.MAN. Note that the iteration (for any algorithm) was stopped if $\max_{1 \le k \le m} \overline{d}_k - 1 \le 0.000001$, thus guaranteeing that $l(\hat{p}) \le l(p^{\text{iterate}}) + 10^{-6} \times n$ by inequality (8.8). The EM algorithm needs a couple of hundred steps to reach this bound, whereas the VEM identifies the non-zero weights in only a few steps.

$\overline{d}_k$	$\hat{p}_k$	k
1	0.0463	1
1	0.0334	2
1	0.0887	3
1	0.0708	4
0.4722	0.	5
0.8337	0.	6
1	0.0926	7
0.7965	0.	8
1	0.0818	9
0.7713	0.	10
0.9377	0.	11
1	0.1209	12
0.9394	0.	13
1	0.4656	14

Table 8.4. Weights and gradient at maximum likelihood estimate
$(\max_{1 \le k \le m} \overline{d}_k \le 1 + 0.000001)$; $l(\hat{p}) = -58.06002$

The estimate given in Table 8.4 corresponds to the one given by Gentleman and Geyer (1994), as it should. However, it is interesting to observe the deviation in the fourth digit for weight 12 and weight 14, which is given as $p_{12} = 0.1206$ and $p_{14} = 0.4658$ by Gentleman and Geyer (1994). This is probably due to the fact that the EM algorithm was stopped too early (we reached a similar value when using the stopping rule $\max_{1 \le k \le m} \overline{d}_k \le 0.0001$), a phenomenon observable quite frequently in mixture models (Titterington, Smith, and Makov 1985, p. 90).

8.4   Estimation of a prevalence rate under herd heterogeneity for the example of bovine trypanosomiasis in Uganda

In prevalence studies the parameter of interest is usually the *prevalence rate* λ, a number between 0 and 1 which if multiplied by 100 can be interpreted as the percentage of infected humans or animals. The prevalence rate is again usually determined by determining a sample size m and the number of infected x out of m is determined leading to an estimate $\hat{\lambda} = x/m$ for the prevalence rate. It follows from the conventional formulas that the variance of this estimate $\hat{\lambda}$ is given by $\text{Var}(\hat{\lambda}) = \lambda(1 - \lambda)/m$ which again can be estimated by $x/m^2(1 - x/m)$.

In the veterinary sciences as one example, however, this procedure is often not completely adequate, because the animal population occurs in herds, and the effect of this is called the *clustering effect*. The sampling takes this into account by sampling from n herds or farms with potentially different sample sizes m_i, $i = 1, ..., n$. The number of infected animals is denoted by x_i for herd or farm $i = 1, ..., n$.

It is common practice in epidemiology to use the *pooled* estimator $\hat{\lambda}_{\text{pool}} = (x_1 + ... + x_n)/(m_1 + ... + m_n)$ as an estimate of the common *prevalence* rate λ. The variance of $\hat{\lambda}_{\text{pool}}$ is readily provided as $\text{Var}(\hat{\lambda}_{\text{pool}}) = ((1 - \lambda)\lambda m_1 + ... + (1 - \lambda)\lambda m_n)/(m_1 + ... + m_n)^2 = (1 - \lambda)\lambda/M$, with $M = m_1 + ... + m_n$. Note that this variance is the identical formula for the variance as in the unstratified sampling.

However, problems occur if the clustering effect cannot be ignored, that is if a common prevalence rate cannot be assumed. Instead, a *heterogeneous* herd population is more likely to be the case with possible different prevalence parameters λ. Then, it was shown in Chapter 6 in eq. (6.1) that the variance of $\hat{\lambda}_{\text{pool}}$ is inflated by a term corresponding to the variance of the population prevalence rate λ. This variance is denoted by τ^2. In formula,

$$\text{Var}(\hat{\lambda}_{\text{pool}}) = \mu(1 - \mu)/M + \tau^2 [m_1(m_1 - 1) + ... + m_n(m_n - 1)]/M^2.$$

Here μ is the overall prevalence rate (the mean of the population prevalence rates) and τ^2 the variance of the population prevalence rate (μ is the expected value and τ^2 is the variance w.r.t. to the heterogeneity distribution P). The above formula demonstrates clearly that if population heterogeneity is ignored the variance of the prevalence estimator is underestimated by the term $\tau^2 (m_1(m_1 - 1) +$

... $+ m_n(m_n - 1))/M^2$. Also, if there is population *homogeneity* ($\tau^2 = 0$) both approaches and formulae coincide. The cluster effect results in a variance inflation leading to increased confidence limits (McDermott *et al.* 1994; Brier 1980; Donner 1993; McDermott and Schukken 1994). To demonstrate these ideas we look at a data set on herd infection with trypanosomiasis discussed in Böhning and Greiner (1998). The data were collected in Mukono County which is located in the south eastern part of Uganda and covers an area of approximately 200 km². The data (listed in Table 8.5) stem from a cross-sectional pilot study launched and accomplished in June/July 1994

Farm	Cases	Sample size	Infection rate	Farm	Cases	Sample size	Infection rate
1	4	9	0.44	26	1	7	0.14
2	0	5	0	27	1	3	0.33
3	3	9	0.33	28	1	11	0 9
4	14	32	0.44	29	1	3	0.33
5	2	17	0.12	30	1	3	0.33
6	0	3	0	31	1	9	0.11
7	1	4	0.25	32	4	9	0.44
8	3	17	0.18	33	0	9	0
9	0	7	0	34	0	7	0
10	0	15	0	35	3	19	0.16
11	0	8	0	36	1	13	0.8
12	0	12	0	37	0	12	0
13	0	9	0	38	5	18	0.28
14	0	16	0	39	2	11	0.18
15	6	16	0.38	40	0	12	0
16	2	5	0.40	41	0	2	0
17	0	9	0	42	2	7	0.29
18	0	6	0	43	2	7	0.29
19	2	8	0.25	44	4	10	0.40
20	0	6	0	45	3	10	0.30
21	0	3	0	46	1	3	0.33
22	1	7	0.14	47	1	15	0.7
23	1	8	0.13	48	0	6	0
24	0	10	0	49	1	6	0.17
25	12	28	0.43	50	1	6	0.17

Table 8.5. Number of cattle infected with Trypanosoma spp., sample size and infection rate for 50 dairy farms in Mukono County, Uganda (data from June 1994, total sample size 487)

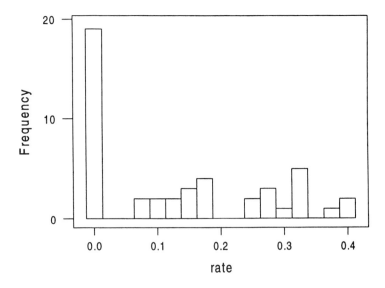

Figure 8.6. Distribution of the prevalence rate of trypanosomiasis over 50 farms.

for a project on trypanocide resistance in the peri-urban dairy production near Kampala. The sampling frame consisted of 187 dairy
farms existing in the region (information from census April 1994)
from which 50 farms were selected at random using random number
tables, stratified for three categories of herd size: small (1–10 cattle),
medium (11–30), large (more than 30). A total of 487 cattle were
sampled on the identified farms.

Frequently, the population heterogeneity is so striking that simple graphical methods are already successful, as in this case. We
consider Figure 8.6 where we provide a histogram of the distribution
of the farm prevalence rate. It is evident from Figure 8.6 that not
one center is present in the distribution but about 3, one at zero, the
second at 0.15, and the third at 0.35. This indicates the presence of
population heterogeneity due to clustering. Estimation of heterogeneity can be done again with C.A.MAN. The mixture model corresponding to (1.1) is

$$f(x_i, m_i, P) = \sum_{j=1}^{k} f(x_i, m_i, \lambda_j)p_j$$

with the binomial mixture kernel $f(x, m, \lambda) = \binom{m}{x} \lambda^x (1-\lambda)^{m-x}$. C.A.MAN provides a heterogeneity estimate with $\hat{k} = 3$ components, as tabulated in Table 8.6.

$\hat{\lambda}_j$	$\hat{p}_j$	j
0.	0.1690	1
0.1161	0.4753	2
0.3195	0.3557	3

Table 8.6. Estimated heterogeneity distribution $\hat{P}$ for prevalence rate of trypanosomiasis

The heterogeneity analysis results in *three* subpopulations: 17% of the herds are infection-free, 48% have an infection rate of 12%, and 36% of the herds show an infection rate of 32%. From this heterogeneity distribution, mean and variances can easily be calculated, leading to

$$\hat{\mu} = \hat{\lambda}_1\hat{p}_1 + \hat{\lambda}_2\hat{p}_2 + \hat{\lambda}_3\hat{p}_3 \text{ and } \hat{\tau}^2 = \hat{p}_1(\hat{\lambda}_1 - \hat{\mu})^2 + \hat{p}_2(\hat{\lambda}_2 - \hat{\mu})^2 + \hat{p}_3(\hat{\lambda}_3 - \hat{\mu})^2$$

It turns out for this data set that incorporating the heterogeneity leads to a variance for the pooled estimator $\hat{\lambda}_{pool} = (x_1 + \dots + x_n)/(m_1 + \dots + m_n)$ about *twice as large* as the variance in the case where homogeneity is assumed. The difference between both approaches is visualized in Figure 8.7. Note that the left confidence interval uses the variance formula $Var(\hat{\lambda}_{pool}) = \lambda(1 - \lambda)/M$, whereas the right is $Var(\hat{\lambda}_{pool}) = \lambda(1 - \lambda)/M + \tau^2[m_1(m_1 - 1) + \dots + m_n(m_n - 1)]/M^2$.

Note that in Chapter 6 a moment estimator of τ^2 has been derived which serves the purpose as well. However, this analysis is more complete in the sense that it not only provides an estimate of τ^2 as a by-product, it also delivers a full estimate of the structure of the population heterogeneity though we have made only partial use of it. It should be briefly noted that the modeling of heterogeneity via the *beta-binomial* distribution (in which the heterogeneity distribution of λ is modeled by a beta-distribution[*]) is known in the literature for the applications at hand (Donald *et al.* 1994; Smith 1983; Madden and Hughes 1994).

---

[*] See also Example 1.5 in Chapter 1.

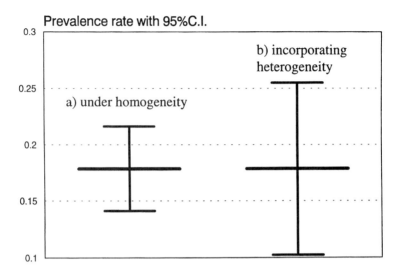

Figure 8.7. Estimation of a prevalence rate with 95% C.I. (a) under assumption of homogeneity and (b) incorporating heterogeneity.

8.5   Random effects or autocorrelation models?

In the previous section the mixture model as a random effects model has been used to model the herd cluster structure. For clustered data *autocorrelation* models are frequently suggested, since there is the reasonable idea that there is a higher probability for disease transmittance within a herd than between herds. In this section we try to point out that random effects and autocorrelation models lead often to the identical marginal model. For binary data we can follow the clear presentation in Collett (1991). Let x denote the number of diseased out of m at risk. Let $y_j = 1$ if the jth unit in cluster i is diseased and $y_j = 0$, otherwise. Note that $x = \sum_{j=1}^{m} y_j$. Suppose that $\Pr(Y_j = 1) = p$. Then it follows that $\mathrm{Var}(Y_i) = p(1-p)$, and consequently $\mathrm{E}(X) = mp$ and

$$
\begin{aligned}
\mathrm{Var}(X) = \mathrm{Var}\!\left(\sum_{j=1}^{m} Y_j\right) &= \sum_{j=1}^{m} \mathrm{Var}(Y_j) + \sum_{j=1}^{m}\sum_{l\neq j}^{m} \mathrm{Cov}(Y_j;Y_l) \\
&= mp(1-p) + \sum_{j=1}^{m}\sum_{l\neq j}^{m} \mathrm{Cov}(Y_j;Y_l).
\end{aligned}
\tag{8.9}
$$

Now, if Y_1, Y_2, ..., Y_m were independent the conventional binomial variance $\text{Var}(X) = mp(1 - p)$ would follow (since all covariance terms are zero). Suppose that there is *autocorrelation*:

$$\rho = \frac{\text{Cov}(Y_j; Y_l)}{\sqrt{\text{Var}(Y_j)\text{Var}(Y_l)}}$$

with $\rho \neq 0$. Because $\text{Cov}(Y_j, Y_l) = \rho\, p(1 - p)$ it follows from (8.9) that

$$\begin{aligned}\text{Var}(X) &= mp(1-p) + m(m-1)\rho p(1-p) \\ &= mp(1-p)[1 + (m-1)\rho].\end{aligned} \tag{8.10}$$

In the case of *positive* autocorrelation ($\rho > 0$) we yield an *over-dispersion* model and the variance inflation is given by $1 + (m - 1)\rho$.

In Chapter 1 a special mixture model was considered for binary data, namely, the beta-binomial mixture model. In Example 1.5, it was shown that if the parameter λ of the binomial distribution follows a beta distribution, then the associated marginal distribution is the beta-binomial distribution. Moreover, it was shown that

$$\text{Var}(X) = m\mu(1 - \mu)\,(1 + \rho(m - 1)) \tag{8.11}$$

with $\mu = 1/(\alpha + \beta)$ being the mean of the beta distribution and $\rho = \theta/(1 + \theta)$, $\theta = 1/(\alpha + \beta)$, α, β being the two parameters of the beta distribution. (8.11) is identical to (8.10) if p is identified with μ. Given this identity it is not astonishing that some authors use the beta-binomial model as an autocorrelation model (though the conventional derivation is by means of a random effects model), nor is it then surprising that it will be difficult to decide solely on statistical grounds if we have an auto-correlation *or* heterogeneity process in population in progress.

8.6   Binary regression and nonparametric mixture models

Frequently in medicine and epidemiology, sociology, economy and other applied sciences the situation of *binary regression* is of interest. There is a dependent binary variable Y, for which $Y = 1$ might represent a diseased person and $Y = 0$ a healthy person, and a vector of x covariates. Then we are interested in the relationship of $p_x = P\,(Y = 1 \,|\, x)$ to the covariates. Often a transformation of p_x, namely, $\log\,(p_x/(1 - p_x))$, is considered and its *linear* relationship to the vector of covariates investigated. The latter is called the logistic regression model $\log (p_x/(1 - p_x)) = \beta^T x$. However, in many situations, this linear relationship is not only in doubt, it is often unclear how a *major* deviation from it can be detected, at least in those situations where no repeated measurements of y are available for a given covariate combination. Various diagnostic devices have been suggested, including the adaptation of nonparametric regression smoothers. Here we suggest another simple alternative using the nonparametric mixture approach. We have that

$$p_x = \frac{f_1(x)\Pr(Y=1)}{f_1(x)\Pr(Y=1) + f_0(x)(1 - \Pr(Y=1))} = \frac{f_1(x)p}{f_1(x)p + f_0(x)(1-p)},$$

where $\Pr(Y = 1) = p$, $f_1(x)$ is the density for the diseased group and $f_0(x)$ is the density for the healthy group. Taking logarithms we achieve

$$\log (p_x/(1 - p_x)) = \log (f_1(x)/f_0(x)) + \log (p/(1 - p)).$$

Note that on the right-hand side p is independent of x. Estimability of p will depend on the study type. If the study is a cross-sectional or cohort type, the prevalence can be estimated. If the study type is case-control p will be *not* estimable. But most important, the covariate x occurs in the two densities only. Consequently, the analysis should focus on the variation of $\log (f_1(x)/f_0(x))$ with x. A quite similar idea has been used recently in Kelsall and Diggle (1998) for the spatial variation of disease risk. It is obvious that estimates for f_i, $i = 0, 1$ must be available and the suggestion here is to use the nonparametric mixture maximum likelihood estimator to get them. In other words, it is suggested to use $\log \{f_1(x, \hat{P}_1) / f_0(x, \hat{P}_0)\}$ to estimate the covariate dependent part of $\log (p_x/(1 - p_x))$, where $f_i(x, \hat{P}_i)$ is the nonparametric mixture density estimate of the cases ($i = 1$) and the controls ($i = 0$), respectively. This is demonstrated in the following example.

Example 8.3: In a follow-up study of schizophrenic patients (Paquing 1995) on the effect of various potential risk factors on the resubmission probability *within six months after leaving a hospital* 251 patients were studied in Metro Manila (Philippines). For the covariate $x =$ *number of persons in household of patient* the two nonparametric maximum likelihood estimates were found and $\log (p_x/(1 - p_x))$ estimated for each covariate combination. In this situation, covariate

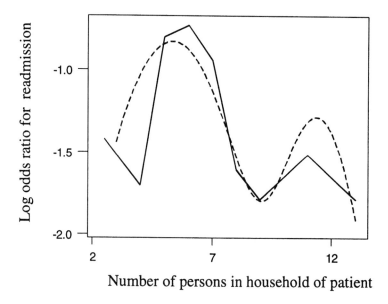

Figure 8.8. Nonparametric binary regression of log odds ratio in dependence of $x =$ *number of persons living in household of patient*; solid curve: empirical logits, dashed curve: estimated by nonparametric mixture model.

combination replicates were available, so that for comparison the estimate $\log[\, f_x/n_x \,/(1 - f_x/n_x)\,]$ was possible to use; here n_x is the number of patients with covariate combination x and f_x is the number of resubmitted patients under the n_x patients with covariate combination x. As can be seen in Figure 8.8 the effect of the covariate x is not only highly non-linear (a simple quadratic term in the linear-logistic would be not sufficient to cover this curvature), the estimate based on the nonparametric mixture model provides a reasonable fit of the nonparametric logits.

8.7   Related topics and open problems

Connection to optimal design theory. Many results in nonparametric mixture models have analogous counterparts in optimal design theory. The latter context can be described as follows. Given a dependent variable Y, a vector of covariates $x = (x_1, \ldots, x_p)^{\mathrm{T}}$ and a connecting linear regression model $\mathrm{E}(Y) = \beta^{\mathrm{T}}x$, based on a sample of size N the best linear unbiased estimator of β is given by $\hat{\beta} = (X^{\mathrm{T}}X)^{-1}X^{\mathrm{T}}Y$, where X is the design matrix

$$X = \begin{pmatrix} x_{11} & x_{12} & \dots & x_{1p} \\ x_{21} & x_{22} & \dots & x_{2p} \\ & \dots & & \\ x_{N1} & x_{N2} & \dots & x_{Np} \end{pmatrix}$$

and $Y = (y_1, \dots, y_N)^T$. The covariance matrix of $\hat{\beta}$ is proportional to $(X^T X)^{-1}$ and $X^T X$ can be written as $N \sum_{i=1}^{N} x_i x_i^T p_i$, with $p_i = 1/N$. In planning an *optimal experiment of size N* one would like to choose those design points which minimize in some sense the covariance matrix $(X^T X)^{-1}$. A frequently used optimality criterion is the determinant, since it measures the contents of the dispersion ellipsoid. In other words, one wants to find that design which maximizes det $(\sum_{i=1}^{N} x_i x_i^T p_i)$, under the restriction $p_i = 1/N$ (exact design). To simplify the optimization task one gives up the restriction $p_i = 1/N$ and maximizes the determinant of $\sum_{i=1}^{k} x_i x_i^T p_i$ under the only constraint $p_i \geq 0$ for all $i = 1, \dots, k$ and $p_1 + \dots + p_k = 1$ (Kiefer 1959). Obviously, one can think of the design as a *mixing distribution* giving weight p_i to the covariate combination x_i for $i = 1, \dots, k$. The results available in optimal design theory can be split into three parts: the *equivalence theory* with the key result in the general equivalence theorem going back to Kiefer and Wolfowitz (1960) stating the equivalence of a continuous design giving mass $p_1, \dots, p_k$ to design points $x_1, \dots, x_k$ which maximizes the determinant and the design which minimizes $\max_x$ $x^T (\sum_{i=1}^{k} x_i x_i^T p_i)^{-1} x$. This result compares closely to the general mixture maximum likelihood theorem (see Atwood 1976, 1980; Titterington 1980; Silvey 1980). The second part is the *duality theory* developed by Silvey and Titterington (1973) and Titterington (1975) (see also Pukelsheim 1993) for which the mixture model counterparts have been pointed out by Lindsay (1983) and Lesperance and Kalbfleisch (1992). The third part of development in optimal design took place in *algorithms*. Most of their mixture model counterparts have been discussed in Chapter 3. Original contributions in the algorithmic area in optimal design are connected with the names of Atwood (1976, 1980), Gribik and Kortanek (1977), Silvey, Titterington, and Torsney (1976), Tsay (1976), Wu (1978, 1983), and Fedorov (1972).

Mixture of normals and unknown variance parameter(s). A few comments for *normal* mixture models are appropriate. Leaving the number of components open *and* the variance parameter unknown (meaning to be estimated) will lead to an *oversmoothed* mixture model.

Either we fix the number of components and then estimate a *common* variance parameter in addition to the means and weights, or we fix the variance parameter and estimate the number of components (as in the other cases with one-parameter mixture kernels). Roeder (1990) discusses an approach in which the *common* variance parameter is constrained and the number of components is estimated. In addition, there is the problem of estimating different variance parameters. Consider the k-component normal mixture

$$\sum_{j=1}^{k} \varphi[(x - \lambda_j)/\sigma_j]p_j$$

where φ is the standard normal. It is well-known that this model was used as an example for failure of the maximum likelihood estimate (see Kiefer and Wolfowitz 1959). Indeed, if $\lambda_j = x$ for some j, we have that $\exp(0)/(2\pi\sigma_j^2)^{1/2} = 1/(2\pi\sigma_j^2)^{1/2}$ which goes to $+\infty$ for σ_j going to zero. In conclusion, for every data point the likelihood equals $+$ infinity. Despite this fact, there exists a version of the EM algorithm providing some kind of *stationary* point (local maximum) at convergence. Though this solution is used in practice frequently, no profound study — at least on the basis of simulation — of the properties of this estimator exists (to the best of the author's knowledge). Another approach was suggested by Hathaway (1985, 1986), essentially constraining the domain of variation of the component variances; consistency of the constrained maximum likelihood estimator could be established and a version of the EM algorithm for computing it could be developed.

Latent class analysis. One of the few developments in mixtures of multivariate distributions has taken place in an area called *latent class models* (Rost and Langeheine 1997). Suppose a number m of binary* items or symptoms $y_1, ..., y_m$ is observed and let $\Pr(Y = y)$ denote the probability that a specific item combination is observed. The *axiom of local independence* claims the existence of k classes such that given the allocation of the unit to class j, $1 \le j \le k$, the joint probability is the product of the marginal probabilities:

$$\Pr(Y = y|\text{class } j) = \theta_{1j}^{y_1}(1 - \theta_{1j})^{1-y_1} \times ... \times \theta_{mj}^{y_m}(1 - \theta_{mj})^{1-y_m}, \quad (8.12)$$

where θ_{ij} is the probability that item i is positive given membership of class j. The idea is that there is a *hidden* covariate such that given

---

* Y is called *binary* if and only if it can take on either $y = 1$ or $y = 0$.

this covariate there is only random residual covariation of the items. The idea goes back to Lazarsfeld and Henry (1968) and has been further exploited and developed by a number of researchers including Clogg (1979) and Formann (1982). The mixture model comes in by considering the *population related* or *unconditional* probability

$$\Pr(Y = y) = \sum_{j=1}^{k} \theta_{1j}^{y_1}(1 - \theta_{ij})^{1-y_1} \times \ldots \times \theta_{mj}^{y_m}(1 - \theta_{mj})^{1-y_m} p_j \quad (8.13)$$

where p_1, p_2, ... , p_k are the population proportions of the k classes. (8.13) shows that we have achieved a very special mixture model. There are a number of problems with this approach, including multiple modes of the likelihood function and difficulties in deciding about identifiability (Goodman 1974; McHugh 1956). For a more detailed discussion the interested reader is pointed to Everitt (1984, p. 76) or Rost and Langeheine (1997).

Working with C.A.MAN

The package C.A.MAN can be downloaded from the author's homepage: http://www.medizin.fu-berlin.de/sozmed/bo1.html.

In the following we provide some simple examples which demonstrate how to work with C.A.MAN. The characters in the text have meanings as follows:

C.A.MAN – output: standard text font

Data-Input of user: **bold face**

Comments of author: *italics*

The main purpose of the package is to compute the Nonparametric Maximum Likelihood Estimator (NPMLE) for a family of densities.

 1: INPUT DATA
 2: CHOICE OF PARAMETER GRID
 3: CHOICE OF DISTRIBUTION
 4: CHOICE OF ALGORITHM
 5: CHOICE OF STEPSIZE
 6: CHOICE OF ACCURACY
 7: COMPUTE NPMLE
 8: GRAPHICS
 9: HELP
10: SPECIAL OPTIONS
11: QUIT

Please enter your choice: **1**

 1: Specify data-file (FIRST THING TO DO)
     NOW: Please specify the structure of your file:
 2: Do you have ungrouped data?
 3: Do you have grouped data of the form: ?
     VARIABLE REPLICATION FACTOR
     Do you want to analyze the mixture of rates or
     SMRs with supposed POISSON (DEFAULT!!)

or BINOMIAL data grouped in the following way:

4:    NUMBER OF EVENTS NUMBER OF EXPOSED (rates)
      OBSERVED CASES EXPECTED CASES (smrs)

5:    NUMBER OF EVENTS NUMBER OF EXPOSED
          REPLICATION FACTOR
          OBSERVED CASES EXPECTED CASES
          REPLICATION FACTOR

6:    Do you want to analyse the mixture of supposed normal-
      distributed data with corresponding variance of the form:
      VARIABLE VARIANCE

7:    Do you want to read a predefined density matrix?

8:    Back to Main Menu

Please enter your choice: **1**

Please enter the name of your data-file:**data1.dat**
This data set was discussed in Example 1.2.

Please enter name of output-file!
If you press <ENTER> the filename will be:

data1.out *In this file the results of the analysis are saved.*

NOW: Please specify the structure of your file:

2:    Do you have ungrouped data?

3:    Do you have grouped data of the form: ?
      VARIABLE REPLICATION FACTOR

Do you want to analyze the mixture of rates or SMRs with
supposingly POISSON (DEFAULT!!) or BINOMIAL data
grouped in the following way:

4:    NUMBER OF EVENTS NUMBER OF EXPOSED (rates)
      OBSERVED CASES EXPECTED CASES (smrs)

5:    NUMBER OF EVENTS NUMBER OF EXPOSED
      REPLICATION FACTOR
      OBSERVED CASES EXPECTED CASES
      REPLICATION FACTOR

6:    Do you want to analyse the mixture of supposed normal-
      distributed data with corresponding variance of the form:
      VARIABLE VARIANCE

7:   Do you want to read a predefined density matrix?
8:   Back to Main Menu
Please enter your choice: **3** *Grouped version of Data Input*
      .00000 120.00000
      1.00000 64.00000
      2.00000 69.00000
      3.00000 72.00000
      4.00000 54.00000

      ... ...
      ... ...

Your count of different variables is: 24

120 Children do have 0 Symptoms, 64 have 1 Symptom, 69 2 symptoms, ...

Choose parameter !

1:   Default parameter grid ! If your sample is less or equal to
     25 every data point is used as a parameter. Otherwise the
     number of grid points is estimated from your data and the
     grid is computed automatically

2:   If you want to compute your grid of parameters individually

3:   If you use fixed support size and if you want to enter
     starting values

4: Back to Main Menu

*Here, an approximation of the infinite probability simplex of all distri-
butions on the parameter space is determined. Default value is the grid
of those values which have been observed in the sample.*

Please enter your choice:

For the following distributions the NPMLE may be computed:

1:   Normal-distribution with single variance
2:   Poisson-distribution
3:   Laplace-distribution
4:   Exponential-distribution

5:   Poisson-distribution for rates
6:   Normal-distribution with different variance
7:   Predefined density matrix
8:   Binomial-distribution

9: Back to MAIN MENU

Please enter your choice: **2**

Here the kernel of the mixture is chosen. Possible are various versions of the normal distribution, Poisson, binomial, Poisson for rates and Observed/expected Ratios, exponential, as well as from a user defined and specified but otherwise arbitrary density matrix.

1:   INPUT DATA
2:   CHOICE OF PARAMETER GRID
3:   CHOICE OF DISTRIBUTION
4:   CHOICE OF ALGORITHM
5:   CHOICE OF STEPSIZE
6:   CHOICE OF ACCURACY
7:   COMPUTE NPMLE
8:   GRAPHICS
9:   HELP
10: SPECIAL OPTIONS
11: QUIT

These are the minimum requirements for the user to set!

Please enter your choice: **7**
PARAMETER-GRID:DEFAULT
DISTRIBUTION:POISSON
ALGORITHM: VEM
STEP-LENGTH: FULL NR
ACCURACY: .000010
NUMBER OF ITERATIONS: 5000
DO YOU WANT SOME CHANGES ? TYPE Y OR N

step 1800 max. dir. derivative 1.000026

After having reached the desired level of accuracy the program has computed the following results :

step 1836 max. dir. derivative 1.000009

The general mixture maximum likelihood theorem is used for defining a stopping rule: the iteration is stopped if the maximum directional Gateaux derivative is smaller than the accuracy chosen by the user!

Log-Likelihood at iterate : −1553.99800

C.A.MAN did identify 8 grid points with positive support

weight: .1405 parameter: .000000
weight: .0943 parameter: 1.000000
weight: .0209 parameter: 2.000000
weight: .4266 parameter: 3.000000
weight: .2064 parameter: 8.000000
weight: .0584 parameter: 9.000000
weight: .0447 parameter: 16.000000
weight: .0083 parameter: 17.000000

NPMLE for default grid approximation of the parameter space corresponding to sample ! Now the computation is finalized using the EM algorithm !

Do you want to compute a refined solution of the NPMLE using the EM-algorithm?

Please TYPE Y (ES) OR N (O)

After the desired number of steps the C.A.MAN has computed the following results:
step 5001 max. dir. derivative 1.000030

group: 1 weight: .0187 mean: .000000
group: 2 weight: .1786 mean: .161550
group: 3 weight: .0208 mean: 2.818795
group: 4 weight: .4589 mean: 2.818795
group: 5 weight: .2105 mean: 8.164904
group: 6 weight: .0587 mean: 8.164904
group: 7 weight: .0455 mean: 16.156230
group: 8 weight: .0084 mean: 16.156230

Do you want to combine identical parameters?

Parameters are considered equal if their difference is smaller than current level of accuracy: .0000100

Please TYPE Y (ES) OR N (O)

y

The NPMLE consists of 5 support points
Result after combining equal estimates:

weight: .0187 parameter: .000000
weight: .1786 parameter: .161550
weight: .4796 parameter: 2.818795
weight: .2692 parameter: 8.164904
weight: .0538 parameter: 16.156230

Log-likelihood at iterate : −1553.81300

Note that there is only a small difference to the likelihood based on the computation of the approximating grid!
    There is also the possibility of visualizing the mixture model graphically in Sub-Menu 6!
    C.A.MAN can also work with a fixed number of components, instead of estimating it! To accomplish this choose Sub-menu 2:

Choose parameter !

1:   Default parameter grid ! If your sample is less or equal to 25 every data point is used as a parameter. Otherwise the number of grid points is estimated from your data and the grid is computed automatically

2:   If you want to compute your grid of parameters individually

3:   If you use fixed support size and if you want to enter starting values

4:   Back to Main Menu

Please enter your choice: **3**

Here one can choose the number of components and starting values! In Sub-Menu 3 choose 5!

You may use the following algorithms to compute the NPMLE:

1:   EM-Algorithm
2:   VEM-Algorithm
3:   VEM-EM (combination of em- and vem-algorithm)
4:   VDM-Algorithm

5:   FIXED SUPPORT SIZE *using EM algorithm*

6:   Back to Main menu

Please enter the number of your choice: **5**

End of Example

There are a couple of data sets in the mixture model literature, where there is debate about which is the right number of components and what are the right MLEs. One of these data sets is the Hasselblad-data of death notice in the *TIMES* in 1910–1913 (see for example, Titterington, Smith, and Makov 1985) found under hassel.dat on the disk. Here is the correct solution:

The NPMLE consists of 3 support points *(one is responsible for the zero inflation)*

Result after combining equal estimates:

weight: .0067 parameter: .000000
weight: .3895 parameter: 1.354982
weight: .6038 parameter: 2.698013

Log-Likelihood at iterate : −1989.92700

Another data set is the Symons-Data on SIDS in North Carolina Counties (see Example 7.4). Originally a two-component mixture model was suggested. Here is the correct solution (use option 4 under INPUT DATA):

The NPMLE consists of 4 support points
Result after combining equal estimates:

weight: .3263 parameter: .001254
weight: .5124 parameter: .002081
weight: .1505 parameter: .003747
weight: .0108 parameter: .009013

Log-Likelihood at iterate : −233.40030

Note that in this case you can work either with the Poisson for Rates or the Binomial mixture kernel [under menu DISTRUBUTIONS option 5 (Poisson distribution for rates) or option 8 (Binomial)]!

Also, there are a variety of utilities available in C.A.MAN under Special Options including classification of each observation into its component using Bayes' theorem or a cut-off value (the latter only for 2-component mixture model).

A few comments for *normal* mixture models are appropriate. Leaving the number of components open *and* the variance parameter unknown (meaning to be estimated) will lead to an *oversmoothed* mixture model. Either we fix the number of components and then estimate a *common* variance parameter in addition to the means and weights, or we fix the variance parameter and estimate the number of components (as in the other cases with one-parameter mixture kernels). This is practical if we have prior knowledge on the variance parameter σ^2. If this is not the case, we suggest the following practical solution. First, σ^2 is estimated as the empirical variance (it can be expected to be an overestimate of σ^2). Then fix σ^2 to this value and estimate the number of components (the default procedure in C.A.MAN). Having found the number of components, we keep k fixed and reestimate the means, weights, and common σ^2.

C.A.MAN can only handle up to 500 *different* observations.* Note that for most discrete distributions the number of different observations is quite limited. Using the replication option, very large sample sizes can be handled (as has been demonstrated in many examples in this book). So the limitation is only a problem for continuous quantities. In that case we recommend grouping the observations into very small intervals and using the replication option. For example, if we

---

* This limitation holds only for the DOS-version of C.A.MAN. There exists also a UNIX-version of C.A.MAN (which can be downloaded from the same web page) which can handle up to 100000 observations and 400 mixture components.

want to analyze systolic blood pressure data, it is usually no restriction to assume that the range is between 1 and 300, and to group the data within units of 1.

References

Agha, M. and Ibrahim, M.T. (1984). Maximum likelihood estimation of mixtures of distributions. *Applied Statistics* **33**, 327–332.

Ahlboom, A. (1993). *Biostatistics for Epidemiologists*. Boca Raton: Lewis Publishers.

Aitkin, M.A. (1995). NPML estimation of the mixing distribution in general statistical models with unobserved random variation, in *Statistical Modelling* (eds. G.U.H. Seeber, B.J. Francis, R. Hatzinger, G. Steckel-Berger), Berlin: Springer-Verlag, pp. 1–9.

Aitkin, M.A. (1996). A general maximum likelihood analysis of overdispersion in generalized linear models. *Statistics and Computing* **6**, 251–262.

Aitkin, M.A., Anderson, D., Francis, B., and Hinde, J. (1990). *Statistical Modelling in GLIM*. Oxford: Clarendon Press.

Aitkin, M. and Wilson, G.T. (1984). Mixture models, outliers, and the EM algorithm. *Technometrics* **22**, 325–332.

Atwood, C.L. (1976). Convergent design sequences for sufficiently regular optimality criteria. *Annals of Statistics* **4**, 1124–1138.

Atwood, C.L. (1980). Convergent design sequences for sufficiently regular optimality criteria, II: singular case. *Annals of Statistics* **8**, 894–913.

Bagozzi, R. P. (1995). *Advanced Methods of Marketing Research*. Cambridge (Mass.): Blackwell.

Bailar, J.C. (1995). The practice of meta-analysis. *J. Clin. Epidemiology* **48**, 149–157.

Becker, N., Frentzel-Beyme, R., and Wagner, G. (1984). *Atlas of Cancer Mortality in the Federal Republic of Germany, Second completely revised edition*. Berlin: Springer-Verlag.

Becker, N. and Wahrendorf, J. (1997). *Atlas of Cancer Mortality in the Federal Republic of Germany 1981–1990*. Berlin: Springer-Verlag.

Begg, C.B. and Berlin, J.A. (1988). Publication bias: a problem in interpreting medical data. *Journal of the Royal Statistical Society A* **151**, 419–463.

Bernardinelli, L. and Montomoli, C. (1992). Empirical Bayes versus fully Bayesian analysis of geographical variation in disease risk. *Statistics in Medicine* **11**, 983–1007.

Bernardinelli, L., Clayton, D., Pascutto, C., Montomoli, C., Ghislardi, M., and Songini, M. (1995). Bayesian analysis of space-time variation in disease risk. *Statistics in Medicine* **14**, 2433–2443.

Besag, J.E. and Mollié. A. (1989). Bayesian mapping of mortality rates. *Bulletin of the International Statistical Institute* **53**, 127–128.

Besag, J.E., York, J., and Mollié, A. (1991). Bayesian image restoration with two applications in spatial statistics (with discussion). *Annals of the Institute of Statistical Mathematics* **43**, 1–59.

Best, N. (1998). Bayesian ecological modelling. *Disease Mapping and Risk Assessment for Public Health,* (eds. A. Lawson, D. Böhning, A. Biggeri, E. Lesaffre, J.-F. Viel, and R. Bertollini), New York: Wiley, ch. 15.

Biggeri, A., Braga, M., and Marchi, M. (1993). Empirical Bayes *interval* estimates: an application to geographical epidemiology. *Journal of the Italian Statistical Society* **3**, 251–267.

Biggeri, A., Divino, F., Frigessi, A., Lawson, A., Böhning, D., and Lesaffre, E. (1998). Introduction to spatial models in ecological analysis, in *Disease Mapping and Risk Assessment for Public Health,* (eds. A. Lawson, D. Böhning, A. Biggeri, E. Lesaffre, J.-F. Viel, and R. Bertollini), New York: Wiley, ch. 14.

Boden, W.E. (1992). Meta-analysis in clinical trial reporting: Has a tool become a weapon? (editorial). *American Journal of Cardiology* **69**, 681–686.

Boeing, H. and Frentzel-Beyme, R. (1991). Regional risk factors for stomach cancer in the FRG. *Environmental Health Perspectives* **94**, 83–89.

Boeing, H., Frentzel-Beyme, R., Berger, M., Berndt, V., Göres, W., Körner, M., Lohmeier, R., Menarcher, A., Männl, H.F.K., Meinhardt, M., Müller, R., Ostermeier, H., Paul, F., Schwemmle, K., Wagner, K.-H., and Wahrendorf, J. (1991). Case-control study on stomach cancer in Germany. *International Journal on Cancer* **47**, 858–864.

Böhning, D. (1982). Convergence of Simar's algorithm for finding the maximum likelihood estimate of a compound Poisson process. *Annals of Statistics* **10**, 1006–1008.

Böhning, D. (1985). Numerical estimation of a probability measure. *Journal of Statistical Planning and Inference* **11**, 57–69.

Böhning, D. (1986). The vertex-exchange-method in D-optimal design theory. *Metrika* **33**, 337–347.

Böhning, D. (1989). Likelihood inference for mixtures: geometrical and other constructions of monotone step-length algorithms. *Biometrika* **76**, 375–383.

Böhning, D. (1991). Grafische Diagnostik unbeobachteter Heterogeneität, in *Gesundheit und Umwelt, 36. Jahrestagung der GMDS* (eds. W. van Eimeren, K. Überla, K. Ulm), Heidelberg: Springer, pp. 72–76.

Böhning, D. (1994). A note on test for Poisson overdispersion. *Biometrika* **81**, 418–419.

Böhning, D. (1995). A review of reliable maximum likelihood algorithms for the semi-parametric mixture maximum likelihood estimator. *Journal of Statistical Planning and Inference* **47**, 5–28.

Böhning, D. and Schelp, F.-P. (1986). A FORTRAN-Subroutine for computing indicators of the nutritional status of children and adolescents. *Statistical Papers* **27**, 141–150.

Böhning, D. and Lindsay, B.G. (1988). Monotonicity of quadratic approximation algorithms. *Annals of the Institute of Statistical Mathematics* **40**, 223–244.

Böhning, D., Schlattmann, P., and Lindsay, B.G. (1992). Computer assisted analysis of mixtures (C.A.MAN): Statistical algorithms. *Biometrics* **48**, 283–303.

Böhning D., Lindsay B., and Schlattmann P. (1992). Statistical methodology for suicide cluster analysis. *American Journal of Epidemiology* **135**, 1310–1314.

Böhning, D., Dietz, E., Schaub, R., Schlattmann, P. and Lindsay, B.G. (1994). The distribution of the likelihood ratio for mixtures of densities from the one-parametric exponential family. *Annals of the Institute of Statistical Mathematics* **46**, 373–388.

Böhning, D. and Suksawasdi Na Ayuthya, R. (1995). Traffic accident mapping in Bangkok metropolis: a case study. *Statistics in Medicine* **14**, 2445–2458.

Böhning. D. and Dietz, E. (1995). Contribution to the discussion of a paper by Traylor and Smith. *Journal of the Royal Statistical Society B* **57**, 33–34.

Böhning, D., Schlattmann, P., and Dietz, E. (1996). Interval Censored Data: A note on the nonparametric maximum likelihood estimator of the distribution function. *Biometrika* **83**, 462–466.

Böhning, D., Dietz, E., and Schlattmann, P. (1998). Recent developments in computer assisted analysis of mixtures (C.A.MAN). *Biometrics* **54**, 367–377.

Böhning, D. and Sarol, J. (1999). A nonparametric estimator of heterogeneity variance with applications to SMR- and rate data. *Biometrical Journal* (submitted).

Böhning, D., Dietz, E., Schlattmann, P., Mendonça, L., Kirchner, U. (1999). The zero-inflated Poisson model and the DMFT-index in dental epidemiology. *Journal of the Royal Statistical Society A* **162**, 195–209.

Böhning, D. and Greiner, M. (1998). Prevalence estimation under heterogeneity in the example of Trypanosomiasis in Uganda. *Preventive Veterinary Medicine* **36**, 11–23.

Böhning, D. and Suksawasdi Na Ayuthya, R. (1999). Analysis of geographical heterogeneity in live-birth ratios in Thailand. *Journal of Biostatistics and Epidemiology,* **4**, 115–122.

Brier, S.S. (1980). Analysis of contingency tables under cluster sampling. *Biometrika* **67**, 591–596.

Carlin, B.P. and Louis, T.A. (1996). *Bayes and Empirical Bayes Methods for Data Analysis.* London: Chapman & Hall.

Cartwright, R.A., Alexander, F.E., McKinney, P.A., and Ricketts, T.J. (1990). *Leukemia and Lymphoma. An Atlas of Distribution within Areas of England and Wales 1984–1988.* London: Leukemia Research Fund.

Chalmers, I. (1993). The Cochrane collaboration: preparing, maintaining, and disseminating systematic reviews of the effects of health care. *Annals of the New York Academy of Sciences* **703**, 156–165.

Cheng, R.C.H. and Traylor, L. (1995). Non-regular maximum likelihood problems (with discussion). *Journal of the Royal Statistical Society B* **57**, 3–44.

Cislaghi, C., Biggeri, A., Braga, M., Lagazio, C., and Marchi, M., (1995). Exploratory tools for disease mapping in geographical epidemiology. *Statistics in Medicine* **14**, 2363–2381.

Clayton, D. and Kaldor, J. (1987). Empirical Bayes estimates for age-standardized relative risks. *Biometrics* **43**, 671–681.

Clayton, D. and Bernardinelli, L. (1992). Bayesian methods for mapping disease risks, in: *Small Area Studies in Geographical and Environmental Epidemiology,* (eds K. Cuzick and P. Elliot), Oxford: Oxford University Press, 205–220.

Clogg, C.C. (1979). Some latent structure models for the analysis of Likert-type data. *Social Science Research* **8**, 287–301.

Collett, D. (1991). *Modelling Binary Data.* London: Chapman & Hall.

Cook, D.J., Sackett, D.L., and Spitzer, W.O. (1995). Methodologic guidelines for systematic reviews of randomized control trials in health care from the Potsdam consultation on meta-analysis. *Journal of Clinical Epidemiology* **48**, 167–171.

Cornwall, J.M. and Ladd, R.T. (1993). Power and accuracy of the Schmidt and Hunter meta-analytic procedures. *Educational and Psychological Measurement* **53**, 877–895.

Cox, D.R. and Hinkley, D.V. (1974). *Theoretical Statistics.* London: Chapman & Hall.

Cressie, N.A.C. (1993). *Statistics for Spatial Data,* Second edition. New York: Wiley.

Cressie, N.A.C. and Chan, N.H. (1989). Spatial modeling of regional variables. *Journal of the American Statistical Association* **84**, 393–401.

Dean, A.G., Dean, J.A., Colombier, D., Brendel, K.A., Smith, D.C., Burton, A.H., Dicker, R.C., Sullivan, K., Fagan, R.F., and Arner, T.G. (1994). *Epi Info, version 6: a word processing, data base, and statistics program for epidemiology on microcomputers.* Atlanta, Georgia, USA: Centers for Disease Control and Prevention.

Dear, K.B.G. and Begg, C.B. (1992). An approach for assessing publication bias prior to performing a meta-analysis. *Statistical Science* **7**, 237–245.

Dempster, A.P., Laird, N.M., and Rubin, D.B. (1977). Maximum likelihood estimation from incomplete data via the EM algorithm (with discussion). *Journal of the Royal Statistical Society B* **39**, 1–38.

DerSimonian, R. (1986). Algorithm AS 221: Maximum likelihood estimation of a mixing distribution. *Applied Statistics* **35**, 302–309, corrected (1990) **39**, 176.

DerSimonian, R. and Laird, N. (1986). Meta-analysis in clinical trials. *Controlled Clinical Trials* **7**:177–188.

Devine, O.J. and Louis, T.A. (1994). A constrained empirical Bayes estimator for incidence rates in areas with small populations. *Statistics in Medicine* **13**, 1119–1133.

Dickersin, K. and Berlin. J.A. (1992). Meta-analysis: state-of-the-science. *Epidemiologic Reviews* **14**, 154–176.

Dietz, E. (1992). Estimation of heterogeneity — A GLM-approach, in *Advances in GLIM and Statistical Modeling*, (eds. L. Fahrmeir, F. Francis, R. Gilchrist, and G. Tutz), Lecture Notes in Statistics, Berlin: Springer Verlag, 66–72.

Dietz, E. and Böhning, D. (1996). Statistical Inference Based on a General Model of Unobserved Heterogeneity, in *Advances in GLIM and Statistical Modeling*, (eds. L. Fahrmeir, F. Francis, R. Gilchrist, and G. Tutz), Lecture Notes in Statistics, Berlin: Springer Verlag, 75–82.

Dietz, E. and Böhning, D. (1997). The use of two-component mixture models with one completely or partly known component. *Computational Statistics* **12**, 219–234.

Doll, R. (1994). The use of meta-analysis in epidemiology: Diet and cancers of the breast and colon. *Nutritional Review* **52**, 233–237.

Donald, A.W., Gardner, I.A., and Wiggins, A.D. (1994). Cut-off points for aggregate herd testing in the presence of disease clustering and correlation of test errors. *Preventive Veterinary Medicine* **19**, 167–187.

Donner, A. (1993). The comparison of proportions in the presence of litter effects. *Preventive Veterinary Medicine* **18**, 17–26.

Eaton, W.W. (1995). Progress in the epidemiology of anxiety disorders. *Epidemiologic Reviews* **17**: 32–38.

Efron, B. and Morris, C. (1973). Stein's estimation rule and its competitors — An empirical Bayes approach. *Journal of the American Statistical Association* **68**, 117–130.

Eggleston, H.G. (1966). *Convexity*. Cambridge: Cambridge University Press.

Eilers, P. (1995). Indirect observations, composite link models, and penalized likelihood, in *Statistical Modelling* (eds. G.U.H. Seeber, B.J. Francis, R. Hatzinger, G. Steckel-Berger), Berlin: Springer-Verlag, pp. 91–98.

Erdfelder, E. (1993). BINOMIX, A BASIC program for maximum likelihood analyses of finite and beta-binomial mixture distributions. *Behaviour Research Methods, Instruments, & Computers* **25**, 416–418.

Everitt, B.S. (1984). *An Introduction to Latent Variable Models*. London: Chapman & Hall.

Everitt, B.S. and Hand, D.J. (1981). *Finite Mixture Distributions*. London: Chapman & Hall.

Fedorov, V.V. (1972). *Theory of Optimal Experiments*. New York: Academic Press.

Fehrer, S.L. and Halliwell, R. E. (1987). Effectiveness of Avon's Skin-So-Soft as a flea repellent on dogs. *Journal of the American Animal Hospital Association* **23**, 217–220.

Feng, Z.D. and McCulloch, C.E. (1992). Statistical inference using maximum likelihood estimation and the generalized likelihood ratio when the true parameter is on the boundary of the parameter space. *Statistics & Probability Letters* **13**, 235–332.

Feng, Z.D. and McCulloch, C.E. (1996). Using Bootstrap likelihood ratios in finite mixture models. *Journal of the Royal Statistical Society B* **58**, 609–617.

Finkelstein, D.M. and Wolfe, R.A. (1985). A semiparametric model for regression analysis of interval censored failure time data. *Biometrics* **41**, 933–945.

Formann, A.K. (1982). Linear logistic latent class analysis. *Biometrical Journal* **24**, 171–190.

Fowlkes, E.B. (1979). Some methods for studying the mixture of two normal (lognormal) distributions. *Journal of the American Statistical Association* **74**, 561–575.

Gail, M. (1978). The analysis of heterogeneity for indirect standardized mortality ratios. *Journal of the Royal Statistical Society A* **141**, 224–34.

Gentleman, R. and Geyer C.J. (1994). Maximum likelihood for interval censored data: Consistency and computation. *Biometrika* **81**, 618–623.

Gibbons, R.D., Clark, D.C., and Fawcett, J. (1990). A statistical method for evaluating suicide clusters and implementing cluster surveillance. *American Journal of Epidemiology* **132**(suppl), S183–91.

Giesbrecht, F.G. and Whitaker, T.B. (1998). Investigations of the problems of assessing aflatoxin levels in peanuts. *Biometrics* **54**, 739–753.

Goffinet, B., Loisel, P., and Laurent, B. (1992). Testing in normal mixture models when the proportions are known. *Biometrika* **79**, 842–846.

Goodman, L.A. (1974). Exploratory latent structure analysis using both identifiable and unidentifiable models. *Biometrika* **61**, 215–231.

Greiner, M., Franke, C.R., Böhning, D., and Schlattmann, P. (1994). Construction of an intrinsic cut-off value for the sero-epidemiological study of Trypanosoma evansi infections in a canine population in Brazil: a new approach towards an unbiased estimation of prevalence. *Acta Tropica* **56**, 97–109.

Greiner, M., Bhat, T.S., Patzelt, R.J., Kakaire, D., Schares, G., Dietz, E., Böhning, D., Zessin, K.H., and Mehlitz, D. (1997). Impact of biological factors on the interpretation of bovine trypanosomiasis serology. *Preventive Veterinary Medicine* **30**, 61–73.

Gribik, P.R. and Kortanek, K.O. (1977). Equivalence theorems and cutting plane algorithms for a class of experimental design problems. *SIAM Journal of Applied Mathematics* **32**, 232–259.

Groeneboom, P. and Wellner, J.A. (1992). *Information Bounds and Nonparametric Maximum Likelihood Estimation*. Basel: Birkhäuser.

Gunga, H.-C., Forson, K., Amegby, N., and Kirsch, K. (1991). Lebensbedingungen und Gesundheitszustand von Berg- und Fabrikarbeitern im torpischen Regenwald von Ghana. *Arbeitsmedizin Sozialmedizin Präaventivmedizin* **26**, 17–25.

Haight, F. (1967). *Handbook of the Poisson Distribution*. New York: Wiley.

Hardy, R. J. and Thompson, S.G. (1998). Detecting and describing heterogeneity in meta-analysis. *Statistics in Medicine* **17**, 841–856.

Harlap, S. and Baras, H. (1984). Conception-waits in fertile women after stopping oral contraceptives. *International Journal of Fertility* **29**, 73–80.

Hartigan, J.A. (1985). A failure of likelihood asymptotics for the mixture model, in *Proc. Berkeley Symp. in Honor of J. Neyman and J. Kiefer* (eds. L. LeCam and R.A. Olshen), vol. II, New York: Wadsworth, pp. 807–810.

Harwell, M. (1997). An empirical study of Hedges's homogeneity test. *Psychological Methods* **2**, 219–231.

Hasselblad, V. (1969). Estimation of finite mixtures of distributions from the exponential family. *Journal of the American Statistical Association* **64**, 1459–1471.

Hasselblad, V. (1994). Meta-analysis in environmental statistics, in *Handbook of Statistics, Vol. 12, Environmental Statistics* (eds. G.P. Patil and C.R. Rao), Amsterdam: North-Holland, pp. 691–716.

Hathaway, R. (1985). A constrained formulation of maximum-likelihood estimation for normal mixture distributions. *Annals of Statistics* **13**, 795–800.

Hathaway, R. (1986). A constrained EM algorithm for univariate normal mixtures. *Journal of Statistical Computation and Simulation* **23**, 211–230.

Haughton, D. (1997). Packages for estimating finite mixtures: a review. *The American Statistician* **51**, 194–205.

Heckman, J. and Singer, B. (1984). A method for minimizing the impact of distributional assumptions in econometric models for duration data. *Econometrica* **52**, 271–320.

Hedges, L.V. (1992). Modeling publication selection effects in meta-analysis. *Statistical Science* **7**, 246–255.

Hedges, L.V. and Olkin, I. (1985). *Statistical Methods for Meta-Analysis.* Orlando, FL: Academic Press.

Hills, M. and Alexander, F. (1989). Statistical methods used in assessing the risk of disease near a source of possible environmental pollution: A review (with discussion). *Journal of the Royal Statistical Society, Series A* **152**, 353–384.

Hinde, J. and Demétrio, C.G.B. (1998). Overdispersion: models and estimation. *Computational Statistics & Data Analysis* **27**, 151–170.

Hoel, P.G. (1943). On indices of dispersion. *Annals of Mathematical Statistics* **14**, 155–162.

Hoffmann, W. and Schlattmann, P. (1998). An analysis of the geographical distribution of leukaemia incidence in the vicinity of a suspected point source – a case study, in *Disease Mapping and Risk Assessment for Public Health,* (eds. A. Lawson, D. Böhning, A. Biggeri, E. Lesaffre, J.-F. Viel, and R. Bertollini), New York: Wiley, ch. 31.

Holland, W.W. (1991). *European Community Atlas of 'Avoidable Death', Vol. 1.* Oxford: Oxford University Press.

Holland, W.W. (1992). *European Community Atlas of 'Avoidable Death', Vol. 2.* Oxford: Oxford University Press.

Jamshidian, M. and Jennrich, R.I. (1997). Acceleration of the EM algorithm by using quasi-Newton methods. *Journal of the Royal Statistical Society B* **59**, 569–88.

Jewell, N.P. (1982). Mixtures of exponential distributions. *Annals of Statistics* **10**, 479–484.

Johnson, N.L., Kotz, S., and Kemp, A.W. (1992). *Univariate Discrete Distributions,* Second edition. New York: Wiley.

Jones, D.R. (1995). Meta-analysis: weighing the evidence. *Statistics in Medicine* **14**, 137–149.

Kelly, A. (1998). Case studies in Bayesian disease mapping for health and health service research in Ireland, in *Disease Mapping and Risk Assessment for Public Health,* (eds. A. Lawson, D. Böhning, A. Biggeri, E. Lesaffre, J.-F. Viel, and R. Bertollini), New York: Wiley, ch. 28.

Kelsall, J.E. and Diggle, P.J. (1998). Spatial variation in risk of disease: a nonparametric approach. *Applied Statistics* **47**, 559–573.

Kiefer, J. (1959). Optimum experimental designs. *Journal of the Royal Statistical Society B* **21**, 272–319.

Kiefer, J. and Wolfowitz, J. (1956). Consistency of the maximum likelihood estimator in the presence of infinitely many incidental parameters. *Annals of Mathematical Statistics* **27**, 887–906.

Kiefer, J. and Wolfowitz, J. (1960). The equivalence of two extremum problems. *Canadian Journal of Mathematics* **12**, 363–366.

Knorr-Held, L. and Raßer, G. (1999). Bayesian detection of clusters and discontinuities in disease maps. *Biometrics* (to appear).

Kowalski, J., Tu, X.M., Day, R.S., and Mendoza-Blanco, J.R. (1997). On the rate of convergence of the ECME algorithm for multiple regression models with *t*-distributed errors. *Biometrika* **84**, 269–283.

Kuan, J., Peck, R.C., and Janke, M.K. (1991). Statistical methods for traffic accident research, in *Proceedings of the 1990 Taipei Symposium in Statistics,* (eds. M.-T. Chao and P.E. Cheng), Taipei: Institute of Statistical Science, pp. 345–395.

Laird, N.M. (1978). Nonparametric maximum likelihood estimation of a mixing distribution. *Journal of the American Statistical Association* **73**, 805–811.

Laird, N.M. (1982). Empirical Bayes estimators using the nonparametric maximum likelihood estimate for the prior. *Journal of Statistical Computation and Simulation* **15**, 211–220.

Lambert, D. (1992). Zero-inflated Poisson regression with an application to defects in manufacturing. *Technometrics* **34**, 1–14.

Last, J.M. (ed.) (1995). *A Dictionary of Epidemiology,* Third edition. Oxford: Oxford University Press.

Lawson, A., Böhning, D., Biggeri, A., Lesaffre, E., Viel, J.-F., and R. Bertollini (eds.) (1998). *Disease Mapping and Risk Assessment for Public Health.* New York: Wiley,

Lawson, A., Böhning, D., Lesaffre, E., Biggeri, A., and Viel, J.-F. (1998). Disease mapping and its uses, in *Disease Mapping and Risk Assessment for Public Health,* (eds. A. Lawson, D. Böhning, A. Biggeri, E. Lesaffre, J.-F. Viel, and R. Bertollini), New York: Wiley, ch. 2.

Lawson, A., Biggeri, A., and Dreassi, E. (1998). Edge effects in disease mapping, in *Disease Mapping and Risk Assessment for Public Health,* (eds. A. Lawson, D. Böhning, A. Biggeri, E. Lesaffre, J.-F. Viel, and R. Bertollini), New York: Wiley, ch. 7.

Lawson, A., Biggeri, A., and Williams, F.L.R. (1998). A review of modelling approaches in health risk assessment around putative sources, in *Disease Mapping and Risk Assessment for Public Health,* (eds. A. Lawson, D. Böhning, A. Biggeri, E. Lesaffre, J.-F. Viel, and R. Bertollini), New York: Wiley, ch. 18.

Lazarsfeld, P.F. and Henry, N.W. (1968). *Latent Structure Analysis.* Boston: Houghton Mifflin Company.

Lesperance, M. and Kalbfleisch, J.D. (1992). An algorithm for computing the nonparametric MLE of a mixing distribution. *Journal of the American Statistical Association* **87**, 120–126.

Levin, R.J. (1987). Human sex pre-selection. *Oxford Reviews of Reproductive Biology,* **9**, 161–191.

Lilienfeld, A.M., Lilienfeld, D.E. (1980). *Foundations of Epidemiology,* 2nd Edition. Oxford: Oxford University Press.

Lindsay, B.G. (1983a). The geometry of mixture likelihoods, part I: a general theory. *Annals of Statistics* **11**, 783–792.

Lindsay, B.G. (1983b). The geometry of mixture likelihoods, part II: the exponential family. *Annals of Statistics* **11**, 86–94.

Lindsay, B.G. and Roeder, K. (1992). Residual diagnostics in the mixture model. *Journal of the American Statistical Association* **87**, 785–795.

Lindsay, B.G. and Roeder, K. (1993). Uniqueness of estimation and identifiability in mixture models. *The Canadian Journal of Statistics* **21**, 139–147.

Lindsay, B.G. (1995). *Mixture models: Theory, Geometry, and Applications.* NSF-CBMS regional conference series in probability and statistics, vol. 5. Hayward: Institute of Statistical Mathematics.

Louis, T.A. (1998). in *Disease Mapping and Risk Assessment for Public Health,* (eds. A. Lawson, D. Böhning, A. Biggeri, E. Lesaffre, J.-F. Viel, and R. Bertollini), New York: Wiley, ch. 5.

Louis, T.A. (1984). Estimating a population of parameter values using Bayes and empirical Bayes methods. *Journal of the American Statistical Association* **79**, 393–398.

Louis, T.A. (1991). Using empirical Bayes methods in biopharmaceutical research. *Statistics in Medicine* **10**, 811–829.

Luoto, R., Kaprio, J., and Uutela, A. (1994). Age at natural menopause and sociodemographic status in Finland. *American Journal of Epidemiology* **139**, 64–76.

MacDonald, P.D.M. (1986). MIX: an interactive program for fitting mixtures of distributions. *The American Statistician* **40**, 53.

MacDonald, P.D.M. and Pitcher, T.J. (1979). Age groups from size-frequency data: a versatile and efficient method for analyzing distribution mixtures. *Journal of the Fisheries Research Board of Canada* **36**, 987–1001.

Madden, L.V. and Hughes, G. (1994). BBD-Computer software for fitting the beta-binomial distribution to disease incidence data. *Plant Disease* **78**, 536–540.

Malone, K.E., Daling, J.R., and Weiss, N.S. (1993). Oral contraceptives in relation to breast cancer. *Epidemiologic Reviews* **15**, 80–97.

Manton, K.E., Woodbury, M.A., Stallard, E., Riggan, W.B., Creason, J.P., and Pellom, A.C. (1989). Empirical Bayes procedures for stabilizing maps of U.S. cancer mortality rates. *Journal of the American Statistical Society* **84**, 637–650.

Maritz, J.S. and Lwin, T. (1989). *Empirical Bayes Methods,* Se- cond edition. London: Chapman & Hall.

Marshall, R.J. (1991). Mapping Disease and Mortality Rates Using Empirical Bayes Estimators. *Applied Statistics* **40**, 283–294.

Martuzzi, M. and Hills, M. (1995). Estimating the degree of heterogeneity between event rates using likelihood. *American Journal of Epidemiology* **141**, 369–74.

Martuzzi, M. and Hills, M. (1998). Estimating the presence and degree of heterogeneity of disease rates, in *Disease Mapping and Risk Assessment for Public Health,* (eds. A. Lawson, D. Böhning, A. Biggeri, E. Lesaffre, J.-F. Viel, and R. Bertollini), New York: Wiley, ch. 26.

McDermott, J.J. and Schukken, Y.H. (1994). A review of methods used for cluster effects in explanatory epidemiological studies of animal populations. *Preventive Veterinary Medicine* **18**, 155–173.

McDermott, J.J., Schukken, Y.H., and Shoukri, M.M. (1994). Study design and analytic methods for data collected from clusters of animals. *Preventive Veterinary Medicine* **18**, 175–191.

McHugh, R.B. (1956). Efficient estimation and local identification in latent class analysis. *Psychometrika* **21**, 331–347.

McLachlan, G. J. (1992). Cluster Analysis and related techniques in medical research. *Statistical Methods in Medical Research* **1**, 27–49.

McLachlan, G.F. and Basford, K.E. (1988). *Mixture Models. Inference and Applications to Clustering.* New York: Marcel Dekker.

McLachlan, G.J. and Krishnan, T. (1997). *The EM Algorithm and Extensions.* New York: Wiley.

Mendell, N.R., Thode, H.C., and Finch, S.J. (1991). The likelihood ratio test for the two-component normal mixture problem: power and sample size analysis. *Biometrics* **47**, 1143–1148.

Meng, X.-L. and van Dyk, D. (1997). The EM algorithm — an old folk-song sung to a fast new tune (with discussion). *Journal of the Royal Statistical Society B* **59**, 511–68.

Mollié, A. (1996). Bayesian mapping of disease, in *Markov Chain Monte Carlo in Practice* (eds. W.R. Gilks, S. Richardson, and D.J. Spiegelhalter), London: Chapman & Hall, 359–379.

Mollié, A. and Richardson, L. (1991). Empirical Bayes estimates of cancer mortality rates using spatial models. *Statistics in Medicine* **10**, 95–112.

NCHS (1976). *National Center for Health Statistics. Growth Charts.* Rockville: US Department of Health, Education and Welfare, Public Health Service, Health Resources Administration.

Neutra, R. (1998). Computer geographic analysis: an editorial on its use and misuse in public health, in *Disease Mapping and Risk Assessment for Public Health,* (eds. A. Lawson, D. Böhning, A. Biggeri, E. Lesaffre, J.-F. Viel, and R. Bertollini), New York: Wiley, ch. 25.

Paquing, D. (1995). *Readmission after Short-Term Psychiatric Hospitalization.* Master-Thesis in fulfillment of the M.Sc.-programme in Epidemiology, University of the Philippines at Manila.

Pendergast, J.F., Gange, S.J., Newton, M.A., Lindstrom, M.J., Palta, M., and Fisher, M.R. (1996). A survey of methods for analyzing clustered binary response data. *International Statistical Review* **64**: 89–118.

Petitti, D.B. (1994). *Meta-Analysis, Decision Analysis, and Cost-Effectiveness Analysis. Synthesis in Medicine.* Oxford: Oxford University Press.

Potthoff, R.F., and Whittinghill M. (1966). Testing for homogeneity. II. The Poisson distribution. *Biometrika* **53**, 183–90.

Pukelsheim, F. (1993). *Optimal Design of Experiments.* London, Wiley.

Rao, C.R. (1989). *Statistics and Truth.* New Delhi: Council of Scientific & Industrial Research.

Richardson, S. and Green, P.J. (1997). On Bayesian analysis of mixtures with an unknown number of components (with discussion). *Journal of the Royal Statistical Society B* **59**, 731–792.

Ridout, M.S. and Morgan, B.J.T. (1991). Modelling digit preference in fecundability studies. *Biometrics* **47**, 1423–1433.

Robbins, H. (1955). An empirical Bayes approach to statistics, in *Proceedings of the third Berkeley Symposium on Mathematical Statistics and Probability* **1**, Berkeley, CA: University of California Press, 157–164.

Roeder, K. (1990). Density estimation with confidence sets exemplified by superclusters and voids in the galaxies. *Journal of the American Statistical Association* **85**, 617–624.

Roeder, K. (1994). A graphical technique for determining the number of components in a mixture of normals. *Journal of the American Statistical Association* **89**, 487–495.

Rosenthal, R. (1994). Parametric measures of effect size, in *The Handbook of Research Synthesis* (eds. H.M. Cooper and L.V. Hedges), New York: Russell Sage Foundation, pp. 231–244.

Rost, J. and Langeheine, R. (1997). A guide through latent structure models for categorical data, in *Applications of Latent Trait and Latent Class Models in the Social Sciences* (eds. J. Rost and R. Langeheine), Münster: Waxmann.

Sánchez-Meca, J., and Marín-Martínez, F. (1997). Homogeneity tests in meta-analysis: a Monte Carlo comparison of statistical power and Type I error. *Quality & Quantity* **31**, 385–399.

Schlattmann, P. and Böhning, D. (1993). Mixture models and disease mapping. *Statistics in Medicine* **12**, 943–50.

Schlattmann, P., Dietz, E., and Böhning, D. (1996). Covariate adjusted mixture models with the program DismapWin. *Statistics in Medicine* **15**, 919–929.

Schlattmann, P. and Böhning, D. (1997). Contribution to a paper by Richardson and Green. *Journal of the Royal Statistical Society B* **59**, 782–783.

Schlattmann, P. and Böhning, D. (1998). Disease mapping with hidden struc-
tures using mixture models, in *Disease Mapping and Risk Assessment for
Public Health,* (eds. A. Lawson, D. Böhning, A. Biggeri, E. Lesaffre, J.-F.
Viel, and R. Bertollini), New York: Wiley, ch. 5.

Schlattmann, P., Böhning, D., Clark, A., and Lawson, A. (1988). Lung cancer
mortality in women in Germany 1995 – a case study in disease mapping,
in *Disease Mapping and Risk Assessment for Public Health,* (eds. A. Lawson, D. Böhning, A. Biggeri, E. Lesaffre, J.-F. Viel, and R. Bertollini), New
York: Wiley, ch. 31.

Seidel, W., Mosler, K., Alker, M., Ruck, A. (1997). Size and power of likelihood
ratio tests in exponential mixture models based on different implementa-
tions of the EM algorithm. Discussion Papers in Statistics and Quantitative
Economics, No. 79, Bundeswehr-University Hamburg.

Seidel, W., Mosler, K., Alker, M. (1999). A cautionary note on likelihood ratio
tests in mixture models. *Annals of the Institute of Statistical Mathematics*
(submitted).

Self, S.G. and Liang, K.-Y. (1987). Asymptotic properties of maximum likelihood estimators and likelihood ratio tests under nonstandard conditions.
Journal of the American Statistical Association **82**, 605–610.

Sillero-Arenas, M., Delgado-Rodriguez, M., Rodigues-Canteras, R., Bueno-
Cavanillas, A., and Galvez-Vargas, R. (1992). *Obstetrics and Gynecology*
79, 286–294.

Silvey, S.D. (1980). *Optimal Design.* London: Chapman & Hall.

Silvey, S.D. and Titterington, D.M. (1973). A geometric approach to optimal
design theory. *Biometrika* **60**, 21–32.

Silvey, S.D., Titterington, D.M., and Torsney, B. (1978). An algorithm for optimal designs on finite design space. *Communication in Statistics, Theory
and Methods* **7**, 1379–1389.

Simar, L. (1976). Maximum likelihood estimation of a compound Poisson process. *Annals of Statistics* **4**, 1200–1209.

Smith, D.M. (1983). Maximum-likelihood estimation of the parameters of the
beta-binomial distribution. *Applied Statistics* **32**, 192–204.

Stigler, S. (1990). The 1988 Neyman memorial lecture: a Galtonian perspective
on shrinkage estimators. *Statistical Science* **5**, 147–155.

Symons, M. J., Grimson. R. C., and Yuan, Y. C. (1982). Clustering of rare
events. *Biometrics* **39**, 193–205.

Thode, H. C., Finch, S. J., and Mendell, N. R. (1988). Simulated percentage
points for the null distribution of the likelihood ratio test for a mixture of
two normals. *Biometrics* **44**, 1195–1201.

Thyrion, P. (1960). Contribution à l'étude du bonus pour non sinsitre en assur-
ance automobile. *Astin Bulletin* **1**, 142–162.

Tiago de Oliveira, J. (1965). Some elementary tests of mixtures of discrete
distributions, in *Classical and Contagious Discrete Distributions* (ed. G.P.
Patil), Pergamon: New York, pp. 379–384.

Titterington, D.M. (1975). Optimal design: some geometrical aspects of D-
optimality. *Biometrika* **62**, 313–320.

Titterington, D.M. (1980). Geometric approaches to design of experiment. *Mathematische Operationsforschung und Statistik, Series Statistics* **11**, 151–163.

Titterington, D.M., Smith, A.F.M., and Makov, U.E. (1985). *Statistical Analysis of Finite Mixture Distributions.* New York: Wiley.

Tsay, J.Y. (1976). On the sequential construction of D-optimal designs. *Journal of the American Statistical Association* **71**, 671–674.

Tsutakawa, R.K., Shoop, G.L., and Marienfeld, C.J. (1985). Empirical Bayes estimation of cancer mortality rates. *Statistics in Medicine* **4**, 201–212.

Waller, L.A., Carlin, B.P., Xia, H., and Gelfand, A.E. (1997). Hierarchical spatio-temporal mapping of disease rates. *Journal of the American Statistical Association* **92**, 607–617.

Walter, S.D. and Birnie, S.E. (1991). Mapping mortality and morbidity patterns: an international comparison. *International Journal of Epidemiology* **20**, 678–689.

Waterlow, J.C., Buzina, R., Keller, W., Lane, J.M., Nichaman, M.Z., and Tanner, J.M. (1977). The presentation and use of height and weight data for comparing the nutritional status of groups of children under the age of 10 years. *Bulletin of the World Health Organization* **55**, 489–498.

Weinberg, C.R. and Gladen, B.C. (1986). The beta-geometric distribution applied to comparative fecundability studies. *Biometrics* **42**, 547–560.

Williams, F.L.R., Lawson, A.B., and Lloyd, O.L. (1992). Low sex ratios of births in areas at risk from air pollution from incinerators, as shown by geographical analysis and 3-dimensional mapping. *International Journal of Epidemiology* **21**, 311–319.

Wu, C.F. (1978). Some algorithmic aspects of the theory of optimal designs. *Annals of Statistics* **6**, 1286–1301.

Wu, C.F. (1983). On the convergence properties of the EM algorithm. *Annals of Statistics* **11**, 95–103.

Author Index

Subject Index